Apocryphal Science

Creative Genius
and Modern Heresies

Neil DeRosa

Hamilton Books
an imprint of
UNIVERSITY PRESS OF AMERICA,® INC.
Dallas • Lanham • Boulder • New York • Oxford

Copyright © 2004 by
Hamilton Books
4501 Forbes Boulevard
Suite 200
Lanham, Maryland 20706
UPA Acquisitions Department (301) 459-3366

PO Box 317
Oxford
OX2 9RU, UK

Library of Congress Control Number: 2004105787
ISBN 0-7618-2899-0 (paperback : alk. ppr.)

Contents

Preface: A Dedication

Is the human race, at the end of science, or at the beginning? It's hard to know. We really have nothing to gauge our progress against except ourselves and our own estimation of what's possible. This book is about how a few courageous scientists have answered that question.

John Horgan, who has written on this subject, (see Chapter 15), believes that we are now at the "end of science" because all of the important things knowable are already known, and every significant discovery and invention has already been made. They have been discovered or made by the great scientists of the past and present. He denies the possibility that there are levels of knowledge possible to us which are unimaginable today. Such a futuristic view, he argues, should be thought of as "ironic science," a kind of literary device or sci-fi fantasy with no connection to reality.

One could imagine "super beings" such as those depicted in Star Trek© who possess a knowledge far exceeding anything that now exists. Or one can accept literally the theories of certain writers who feel that "ancient astronauts" possessing such knowledge once walked the Earth. But all of that may be mere speculation—or as Horgan says—ironic science.

On the other hand, it is not unreasonable to suggest that every great discovery, and every new insight is an act of creation; a visionary, human creation of something previously unknown or unimagined. If so, then every new idea is a conceptual leap beyond the present limit, or "end," of our knowledge. But who is to say if today's limit of knowledge is the final or ultimate *end* of it? I suggest that no one is, because their argument would be predicated on the ability to see the future, and know the not-yet-known. On the other hand, if past events are any indication of future possibilities—then it is fair to say that the sky's the limit, that the future of science looks bright.

That is why mankind must remain free to look beyond the present plateau and on to the next horizon. There will always be scientists who will do that, if left unfettered to pursue their goals. The innovative scientists and thinkers discussed in these pages, although unknown to many, are at the leading edge of their various areas of specialization. Not all will be found to have been correct when history renders its final verdict. But what they all have in common is a certain optimism, and a philosophy of science which makes the human search for knowledge possible.

In today's environment it takes courage and determination to follow this course, and integrity not to waiver when pressure is brought to bear. Bucking the tide of the mainstream, coming up against the concrete-bound mindset of the high priests of science, many will falter. But if they succeed, and if they are proven right, they will make the world a better place and we will all benefit.

This book is dedicated to such men and women.

*   *   *

I would especially like to thank the "Principals" of my story for their reading of relevant chapters, for their encouragement and many helpful suggestions—technical, scientific, and otherwise. In this regard I thank the following: Tom Van Flandern, Robert Schoch, Peter Breggin, Peter Duesberg, Halton Arp, Michael Behe, Carver Mead, and Patrick Michaels.

I would also like to acknowledge the many helpful suggestions I have received from readers of this manuscript when it was a work in progress. I thank in this regard, Phyllis Trier and Antoinette D'Ammora. I would like to thank Erika Holzer and Michelle Fram-Cohen for encouraging me in my writing; Tom Bethel who, through his work, pointed me in the right direction more than a few times; Karen Minto, David Kelley, Marsha Enright, Nancy Demetry, David Oyerly, Arnold Baise, Richard DeRosa, Richard Shedenhelm, and Elaine Ring for their helpful information, suggestions and insights, and support—all of which contributed to making this a better book than it might have otherwise been.

Since this work may be considered a controversial treatment of several controversial subjects, I state for the record that although I have endeavored to be factually correct, I take full responsibility for any faults in the final product. Any errors are mine alone.

Neil DeRosa
May, 2004

Introduction:
What is Apocryphal Science?

DEFINING TERMS

One of the first rules we learn when making any case, but a rule rarely followed, is the one which says, "Define your terms." So I'll begin right off by defining them.

The meaning of the word *Apocryphal* intended in the following pages is: "books and the theories they contain that are rejected, or regarded as not canonical, and not recognized as valid by some established authority or hierarchy empowered to render such a judgment." This kind of definition is usually meant to refer to the books of the Bible and not to scientific books; but still the meaning is clear, and will be what I mean when I use the word.

Apocryphal theories in this view challenge canon law, dogma, authoritarianism, and closed minds. They offer instead free thought, new ideas, experimentalism, and open minds. Therefore, in this book, the authors of the challenging new theories which are in opposition to the scientific dogmas of our times will be called *Apocryphal scientists*. The principal scientists we will discuss in these pages, although they are certainly not the only ones who have proposed revolutionary new ideas recently, will be called *Principals*. This is their story, not mine.

This is not a book about religion but about science, and in science, as the old saying goes, the proof of the pudding is in the eating. The truth of a scientific theory is in the validity of the facts, the strength of the evidence, and the logic of the arguments used to support it—meaning the theory must be testable, and falsifiable. In other words, other scientists must be able to take the theory and test it in some empirical way to see if it's right or wrong. "Authority" should have nothing to do with it. But in the real world, authority has a lot to do with it, and canon law is every bit as important in science today as it was in the Medieval Church of olden times. That's why the term *Apocryphal* as a challenge to today's "Orthodoxy," is appropriate. It has the same kind of meaning and justification now as it did then.

TO THE READER

I should make a few things clear at the outset. This book is written for the general intelligent reader and for anyone interested in science. It is about revolutionary new ideas in science from a wide range of scientific disciplines, and is therefore generalist in nature. It is written on the level of popular science and does not get into overly technical details. To discover the details, and the formal proofs, the reader must turn to the principal sources, namely the books and articles written by the Principals of my story, many of which are listed in the References section of this book. The reader should also be made aware from the outset that this book is not anti-mainstream science. Most of mainstream science is obviously valid, and is responsible for all of the good things in life resulting from such science. Yet there are certain trends, trends which may pose a great danger to the future of progress.

ON CONSPIRACY THEORIES

I'll also admit right off that I am a conspiracy theorist, but the good kind! It is my opinion that most conspiracy theories are started and spread by cranks or crooks for the purpose of exploiting people who want something to believe in.

If you follow that kind of thing because it's fun, it's okay as long as you know it's fiction—for entertainment only. But if you take it seriously, there are several supermarket check-out counter tabloids, and many books dedicated to this kind of stuff which may be your cup of tea. Conspiracy theories of this kind are well known, and well publicized. Mainstream authorities, (and the media), are not afraid of them because they are easily refuted. That's why UFO commentaries are frequently seen on TV. They pose no threat to the status quo because they can not be verified or falsified. In short, everyone knows they are not about real matters of fact—everyone except for the gullible.

But the second kind, or class, of "conspiracy theories," the kind I believe in, is different. They are far more dangerous to the powers-that-be. For one thing you rarely ever see or hear about them. Advocates of this type are not cranks, (nor even conspiracy theorist), but they are often portrayed as such. They are either scientists or other specialists in their fields. Obviously, they too are sometimes wrong; but they are feared nonetheless, because, *what if they are right*? It could be disastrous to the status quo, and to the current *paradigm*, (the way things are now seen and done in the world). A lot of powerful people have a lot to lose, so such theories are censored, silenced, suppressed. There is a general blackout on them. But living in a free society, you and I can still find out about them if we make a little effort. This book is one such effort.

ON THE LIGHTER SIDE

Recently, I decided to go on a diet. So I set my sights on the chore ahead. This diet would take about a year, just about enough time to write this book!

There is a right way and a wrong way to do everything, and dieting is no exception. There is no magic way to lose weight; you just have to consume less calories than you burn, and you'll lose. Guaranteed. The trouble is you're always hungry.

I set my goal knowing I could do it, having done it twice before. But since it always helps to get motivated before doing something, I went out to pick up the best diet book available. I knew just the right one, having read it before. But since I had given my copy to a friend, I needed another. So I went to a major chain bookstore to pick it up.

Some things never change. The bookstore, a very large chain store, had three rows of shelves loaded with books on how to lose weight. Eat lots of protein and lose—two shelves; eat lots of pasta or "complex carbs" and lose; meditate and lose; eat grapefruit and lose; don't eat meat at all and lose; and many more like them. I had read a few of them so I knew a little about the subject. One popular dietician even gave serious consideration to the belief held by certain gurus that we can live well and be healthy on no food at all; providing that the proper state of consciousness is achieved.

Of course diet fads are nothing new. There have always been good diets, bad diets and all kinds in-between. There are common sense diets, but mostly there is the other kind, the kind which promise the easy way out. Among these are the diets which promise that you'll lose all the weight you want without dieting at all—provided you take a certain very expensive pill. Then there is the option of losing weight by surgical intervention—cut out the excess pounds.

But having read about the subject, and having thought about it for a long time, I already knew what to do. It's all about a balanced diet, plenty of exercise, and rest, right? Still I needed some help with motivation. So I wanted this book again. It's fairly recent, and I thought, a popular one. But when I went to pick up a copy, I couldn't find one at the bookstore. It wasn't just a matter of being temporarily out of stock. The last few times I had been there browsing they were always out of this particular volume.

Being a conspiracy theorist, I wondered why.

The book is *The Omega Diet*, by Artemis P. Simopoulos, M.D., and Jo Robinson; but it is clear that Dr. Simopoulos is the principal author.

I was able to order a copy, of course, since dieting, even in America, is not a very controversial subject. Still, I wondered why Dr. Simopoulos' work was so under-represented while all the crack-pot theories had free reign. Why was the best always the hardest to find? In a country where knowledge and information is freely available, you would think that the opposite would be true. (See Appendix B for more on the *The Omega Diet*).

Of course, my opinion as to what's best could be completely wrong, completely subjective and biased. But I don't think so, and I'll try to convince the reader of that as we proceed.

THE SUBJECT

My story is not about fad diets or "junk science," but about slightly weightier matters if you'll pardon the pun. It's about a perceived threat to the future of science, and the things that science brings, that we all need, caused by the kind of attitude that I intuitively glimpsed in the case of my favorite diet book. The subject matter of my book encompasses certain recent controversial scientific theories, and some little known ones, and the way the mainstream scientific community has received them—or failed to do so.

The official reason always given for the negative reaction is of course that the theory in question is wrong. It might be wrong on a number of counts. It could be based on faulty scientific reasoning or research. The experiments used to demonstrate a theory could be non-repeatable. New facts might be discovered which disprove the theory. Reactions from the scientific community can vary; from rejecting a new theory on peer review, to refuting it in the scientific journals, to character assassination, ridicule, and ad hominem attacks in the press, to totally ignoring the dissenting theory.

Of course, one must be able to give up on a theory that really is wrong, and this is what separates conspiracy theory from real science. But what of innovative or revolutionary new ideas that have a reasonable claim to viability and credibility, and seem to demonstrate good science and valid methodology? These too are often given short shrift or otherwise dispatched with rudely, especially if they are perceived to be a threat to the status quo by the keepers of what Thomas Kuhn calls "normal science" and what we will refer to in these pages as the "scientific establishment," or "mainstream." Does this mainstream have the ability, or the integrity, to give up its own cherished theories if the facts demand it? Do the scientific facts govern its decision making process, or is that process governed by the social psychology of the controlling authority and the individuals who compose it? I accept Kuhn's thesis that existing paradigms are often adhered to with ferocious tenacity, and that scientific revolutions come about only with great difficulty. But there is more to it than that, other important dynamics which complicate this model and pose a greater danger.

The principle scientists surveyed in these pages have all challenged existing paradigms. Some of them may be wrong also, but some may not be. This is what we will try to find out.

ONE MORE NOTE BEFORE WE BEGIN

Since, as I have said, this book is written for the intelligent lay person, scientists who might read it may find the explanations simplistic. But in order to have a reasonable expectation of validity, it will need to be accurate. Therefore, I will try to represent the arguments as accurately as possible. I will keep my sources and resources to a minimum, but will include enough of them to allow the reader to pursue an independent investigation if he or she chooses to do so. And since mathematics is often part and parcel with science, I'll say that there will be no math here—or almost none. That wouldn't do in a book for the general reader. I ask any scientists or students of science to resist the urge to be condescending on that account, and not to give in to the desire to dismiss the book (or at least certain chapters) out of hand. I ask them to be patient; read the evidence presented with an open mind, and then decide.

The Principals of my story are all experts in their fields and are for the most part themselves mainstream scientists, well respected by their peers—that is until they became heretics by publishing and championing taboo theories. This is a forum for the popular dissemination of their ideas.

My task will be to lay out the theories in brief summaries, try to make them understandable to the reader, and to connect certain other logical dots. I'll explain why I think they are right—or wrong, and sometimes to describe the reaction of the peers in the scientific community and the intellectual community at large. Peer reaction comes to the attention of the general public by way of the media; newspapers, magazines, television, and radio, and also in books popularizing the subjects. Some of the dissenting theories are hardly known, (that's one of the reasons for calling them *Apocryphal*), though the mainstream theories which they oppose are usually very well known to the general educated public. There is a concerted effort to silence the heretics, and if that can't be done to ignore them, and if that can't be done, to discredit them.

Chapter One

The Principals:
Unrecognized Trailblazers of Progress

People in general love phenomena that are quite different from the world
they are used to. What they do not like is being asked to abandon reason.

—Petr Beckmann

SOMETHING'S WRONG WITH TODAY'S SCIENCE

No one can doubt that we live in a modern scientific age. By that I mean that
we live in a world in which machines and products of science are in common
use, and that the theories which make them work are fairly well understood.
It is true that as long as civilization has existed on Earth—and probably before that—humans have always used tools; and tools are really only simple
machines. In this sense then, tools can be thought of as the products of science also, since all machines, even simple ones, imply some level of science.

Although all knowledge, including scientific knowledge, has its roots in antiquity, science in the modern sense is unique. In the modern age machines and
the theories behind them have been developed to a level of sophistication and
complexity which entitle them, in no uncertain terms, to be called products of
science. This is how I use the term here. Think of computers, modern medicine,
electricity and electronics, atomic energy, automation, the Internet, mechanized
industry and agriculture, air travel, and space travel. That's what I mean when I
say that we live in a scientific age characterized by the products of science.

Building on the liberating events of the Renaissance and the Reformation,
this scientific age we live in began with the Industrial Revolution, the Enlightenment, and the Scientific Revolution of a mere three hundred years ago.
It was marked by the development of higher mathematics and an empirical

1

scientific method (Hall 1981). It arose first in the freer societies with the freest markets, in England, France, The United States, Germany, Italy, and elsewhere (Rand 1967; Hayek 1954). From these Western nations it spread to the rest of the world. This scientific age even today seems to be progressing steadily and developing exponentially.

Yet, there is something wrong with science today. There are signs of trouble. The ideal of free enquiry seems to be loosing its vitality. New discoveries and reasonable theories are being withheld or rejected without a hearing. New ideas which can have profound importance for the further advancement, the health, and progress of mankind are being ignored. There are attempts to put over "Big Lies" for political or ideological reasons, which could do even more damage than the truth withheld. I will begin by stating for the record that I hope I'm wrong. But I don't think so—there are too many trends and signs. These indicators should warn us that the methods and very state of mind which make the modern age of science possible are in jeopardy and unless corrective action is taken, science may at the very least be damaged or impaired, or more seriously, be caused to stagnate and decline.

A reasonable objection to what I have just said might be that science is being threatened today in the same way it has always been threatened in the past. New scientific theories have always had difficulties in supplanting older existing models or paradigms (Kuhn 1970). Historically also, there have always been periods of flowering of the arts and sciences followed by periods of decline or stagnation. Are we perhaps simply at the juncture of one of those periods? There are indications that we are. There are symptoms of a change for the worse.

But the key difference, the new factor, seems to be in the growing power of the scientific establishment, which is increasingly linked to government agencies—and ever increasing public, (meaning taxpayer), funding. The essential point is that this state of affairs gives rise to a mainstream establishment with the power to suppress dissent, even well reasoned, scientific dissent. This is a power, which in freer times it did not have. In a way, this establishment resembles the dogmatic, Medieval Church of a thousand years ago. This is a comparison I will make use of occasionally in this book.

Will this entrenched status quo succeed in destroying science, or seriously hampering its progress? The future is never sure. But whether or not that happens, it is important to identify the problem and give details of the process. This we will do as we progress. More importantly, we will bring into light of day the actual science which is being suppressed, in terms, as we have said, that any intelligent person can understand—and let the reader decide its merit.

One more note of explanation: It is interesting that some, though not all, of the Principals under consideration, are unaware of other battles being fought, Apocryphal scientific theories in fields outside of their own areas of expertise. Some of them, though not all, seem not to be aware of the reasons behind this threat to science, or that there may even be a threat as such. It is also interesting to note that they come from all parts of the political spectrum, indicating perhaps that this potential crisis in science is not, or not usually, politically motivated.

THE PRINCIPALS: THEIR THEORIES IN A NUTSHELL— LISTED ALPHABETICALLY:

Halton Arp

The theory: Observed *redshifts* are not the result of receding galaxies moving away from us at ever increasing speeds beginning with the "Big Bang," which is the supposed creation of the universe out of nothing. *Quasars* are not the furthest and brightest objects in the universe. Instead their high redshifts indicate *new matter* and *new galaxies* being formed from older central galaxies in our *local* supercluster of galaxies. The universe as a whole, (though not its individual parts), may be infinite in size and age, or at least much older and much larger than previously thought.

Petr Beckmann

The theory: *Einstein's Special Theory of Relativity is wrong.* Classical physics and *Galilean Electrodynamics* better explain the physical world. Beckmann's theory impacts on the optical evidence, electrodynamics, quantum physics, and gravity. This theory is a first attempt to place Special Relativity on "classical grounds," i.e., to render it obsolete.

Michael Behe

The theory: Life at the molecular level is much too complex to have evolved according to the Darwinian model. Natural selection by random variation and mutation is not the mechanism for evolution where *irreducible complexity* is evident. Life must therefore be the product of *design*. Although the concept of "design" has certain religious implications, this is nevertheless a scientific hypothesis worth considering.

Peter Breggin

The theory: Most psychiatric diseases do not have a biological cause; pharmaceuticals and other treatments are doing immeasurable harm to the very people they purport to help. Traditional methods using love and understanding are more beneficial. Breggin chronicles the dangers of many psychiatric drugs, shock therapy, lobotomy, and much more.

Peter Duesberg

The theory: *AIDS is not caused by the HIV virus.* Medications used to treat AIDS patients are killing them. AIDS is caused by long term, heavy use of *recreational drugs.* This chapter discusses Duesberg's seemingly hopeless battle to stand up against the "AIDS Establishment."

Rhawn Joseph

The theory: Joseph develops further and in detail, evidence supporting the existing idea of "panspermia," which he calls *astrobiology.* He postulates that life did not originate here on Earth but was instead transplanted here by means of space debris soon after the Earth was formed. He also offers a substitute theory to describe biological evolution, thus replacing Darwinism, with a pre-determined *evolutionary metamorphosis*, which is pre-programmed in DNA.

Carver Mead

The theory: Quantum physics can be explained by studying and understanding the collective nature of matter in the coherent systems of *Collective Electrodynamics* as exhibited in superconductors, lasers, and the like. A radical challenge to the "Copenhagen Interpretation" is offered by this otherwise mainstream scientist and Caltech pioneer in computer science.

Patrick Michaels

The theory: *Global warming is not the result of man-made greenhouse gases* and is not occurring according to prediction models. Moreover, if there is some warming, it may be negligible, of natural origin, and beneficial to life. This scientific challenge opposes the near unanimous position of scientists on the political Left.

Zecharia Sitchin

The theory: A borderline Apocryphal science view that the ancient "gods," were really aliens from another planet, who came to Earth in pre-historic times, genetically engineered *Homo sapiens* from lower primates or hominids, in the "gods" image, and gave us the fundamental ingredients of our first civilizations. This might be an outlandish theory, even for Apocryphal science, except that there may be some good evidence to support it.

Robert Schoch

The theory: Advanced, sophisticated civilization began long before, perhaps thousands of years before, the conventionally accepted date of circa, 3500 BC. The primary evidence he offers to support this theory is the geological weathering patterns on the *Great Sphinx* in Egypt. There is also other compelling evidence which will be discussed.

Tom Van Flandern

The theory: There is abundant evidence to support an *exploded planet hypothesis*. Many implications of this hypothesis, for which there is ample evidence include; a new theory of comets, a new explanation for the extinction of the dinosaurs, connections to possible artificial structures on Mars, and more. Van Flandern is also a critic of Einstein's theory of Special Relativity, and some of the bizarre implications of quantum mechanics. He is the founder of a radically new model—a scientific system which he calls *Meta Science*, which includes an epistemological method that leads him to his various conclusions. He is also a leading proponent of a new theory of *pushing gravity*—all of which will be discussed.

OTHER VOICES, NOVELIST/ PHILOSOPHERS, DOOMSAYERS, AND JUNK SCIENCE

There are other subjects of Apocryphal scientific interest which fit into the general scope and theme of this book, for which there may not be one dominant Principal, in the sense of being a scientist who is the founder of a radical new theory capable of replacing the existing model, and for which there is valid scientific evidence and a falsifiable hypothesis. To this category also belong Thomas Kuhn, Tom Bethel, Alastair Rae, Ayn Rand, John Horgan, and various other writers and scientists who have addressed some of the subjects to be discussed on the following pages.

CONSIDERING GALILEO

To the casual reader, some of these theories may seem outrageous at the outset. Some others may seem quite non-controversial. It depends, I suppose, on
the reader's own background and point of view. But take my word for it; all
of the theories are equally beheld with horror and dismay in their respective
fields, and for fundamentally similar reasons. More importantly, all of the theories may be true or scientifically valid. Or if not presently verifiable, they
may at least be credible in some respects, and deserving of further research
and investigation.

The reader may not be convinced from the foregoing that science itself is
in any great danger. All I can say to that is that I hope you are right. But
haven't forbidden ideas always been the ones which were new, unconventional, heretical, and in conflict with the conventional wisdom? Isn't it true
that if *any* theory, which is credible, and reasonably presented, is held up to
ridicule and hounded out of the public forum; that important new ideas can
thus be killed or stillborn in the process? If so, isn't that a prelude to the destruction of science?

For the reader who is still in doubt, consider the famous story of Galileo.
Trying to convince the Churchmen of new truths, he bid them to look at the
moons of Jupiter through his new telescope. Some refused to look, but others
did so. Those who viewed the moons still would not accept Galileo's theories.
It must be some kind of trick, they insisted; for everyone knew that the heavens revolved around the Earth.

Chapter Two

Intelligent Design:
Science or Old Time Religion?

THE BIRTH OF THE SCIENTIFIC VIEW

This chapter is about Michael Behe's theory of intelligent design. It is often dismissed as a religious theory by biologists and others familiar with his work. What I think about it, and why, will be explained in due course. But before I do that, I want to say at the outset that if it is to be considered as a viable scientific theory, it can not be incompatible with the existing factual evidence of evolution, and with the fossil record—notwithstanding that creationists have tried to deny these things, and conversely, that Darwinists have sometimes been known to alter the facts. In order to help the reader decide, I begin with a brief explanation of what science in general means, and why the facts of biological evolution can not be denied.

The idea of design is an old one, originating long before the beginning of recorded history. The archeological record of the ancient Sumerians and Babylonians, indicating their pantheons of deities, is clear. Anthropologists and archeologists have shown, moreover, that long before civilization began, primitive humans had already believed in spiritual things. The fact that they buried their dead seems to be a clear indication of this. Artifacts and symbols buried with them, and paintings found in their caves, also indicate spiritual belief (Leakey 1992). It is reasonable to assume that at least by forty thousand years ago, there was some idea in the minds of men that spirits or gods caused things to happen; caused the seasons to change, the old to die, and the hunt to be successful. They also probably believed that these beings or forces created life. In one form or another then, most people have always believed in design. But the fact of a belief does not make that belief true, and for that reason science exists. Science looks for the logical and material causes of things.

This habit of looking for causes, rather than for spiritual explanations, began well before the birth of modern science. Inquisitive minds have always asked questions. Before modern times, such questioning was the province of philosophers. There have always been those who understood that "the gods" and religion could be explained in scientific terms. In ancient Greece, there were many thinkers who held such beliefs, notably the Epicureans. In Roman times, there was Lucretius. At the dawn of the modern age, there were materialists such as Thomas Hobbes, and advocates of the new experimental science such as Francis Bacon. What they all had in common was that they denied miracles, challenged "idols," and demanded proof.

Finally, during the last three hundred years, beginning with Galileo and Newton, the search for practical truth and logical causes, became the province of science. Facts were discovered, and theories laid out, which couldn't easily be refuted. Astronomy replaced astrology; chemistry replaced alchemy; and modern physics replaced natural philosophy, mathematics expanded into the most exacting realm. The Scientific Revolution had begun (Hall 1981).

In biology, Charles Darwin's "theory of evolution by natural selection," popularized by Herbert Spencer's "survival of the fittest," became the accepted theory of the Western world; but not without a fight. This was because it was a direct challenge to the cherished beliefs of religion—specifically, of Christianity. Nevertheless, during the first half of the twentieth century, the Theory of Evolution became the accepted truth of educated society, the new conventional wisdom. But why?

Because science had said it—and proved it. The Scientific Revolution was in full swing. Although most people still believed in one form of religion or another, the tenets of those religions had lost much of their power to inform scholarly opinion. This was especially true of the story of Creation—the belief that a super intelligent being, God, designed or created the world and man.

One optimistic view of the world, the view I share, states that in a "free market place of ideas" people will usually choose the more reasonable of the alternatives they are offered. In biology, there was Darwin's theory—*evolution by natural selection*—and there were facts, lots of facts. There was the fossil record which could not be ignored. There were scientific dating methods which let us know the age of fossils and the artifacts found. And most important, there was the compellingly logical model offered first by Wallace, and then explained in greater detail by Darwin.

EVOLUTION AT A GLANCE

The fossil record shows the rise of mammals, then of the primates, and then, *Homo sapiens*—man. It even shows the beginnings of life on Earth. Darwin had speculated on a primeval pond where life had sprung from the basic elements of the Earth. Since that first speculation, there has been much research and theorizing. It is now thought that the first life, in the form of some type of bacteria, arose in the hot steamy cauldrons of the forming Earth—around four billion years ago. (Fortey 1997; Dawkins 1996).There has been some evidence to suggest that the process may have been initiated by, or at least aided by, organic matter in meteorites arriving from outer space (Joseph 2001). This is the idea of "panspermia," or astrobiology, and is one of the Apocryphal theories to be discussed in this book, since it is not yet accepted by mainstream science.

From the first beginnings of life, the fossil record can be followed through its various stages of development; including the Cambrian explosion, the Permian extinction and subsequent rebound, and the rise of the vertebrates. We know of the age of the dinosaurs and of their extinction around sixty-five million years ago. We know of the rise of the placental mammals—warm blooded animals with hair and mammary glands, which reproduce their offspring internally. The primates arose from this class of vertebrates, then the great apes, then the hominids, and then finally *Homo sapiens*—"thinking man," or modern man (Fortey 1997).

Fossils of the first modern man, ("modern," in this sense meaning capable of learning or doing whatever you or I can learn or do today—and physically looking like you and me), were found and dated at around forty thousand years old. These were named Cro-Magnon man, after the place in France, in which they were first found. Other fully modern *Homo sapiens* fossils have since been found which are dated at between 100,000 and 150,000 years old. Today, both the first bipedal hominids, which the fossil evidence shows to be four to five million years old and fully modern man, are thought to have originated in Africa (Leakey 1992).

Neanderthal men, another primitive species of almost modern humans, were known to have coexisted in Europe and the Middle East with *Homo sapiens* for millennia, between approximately 100,000 and 32,000 years ago. Neanderthal man is well known because he buried his dead and thereby left a good fossil record. This species is now extinct. It may have been exterminated by Cro-Magnon man, or have died out by natural causes such as the ice age and a dwindling food supply. The scientific jury is still out (Leakey 1992).

Eventually, around ten thousand years ago, modern men and women somehow learned to plant their own food and domesticate animals rather than hunt and gather them in the wild; although this was undoubtedly a gradual process, beginning perhaps many thousands of years earlier. With the beginnings of agriculture came the beginnings of civilization. And so the story goes.

This is the science of it (very briefly and simply expressed). It is proven by the various related sciences, archeology, anthropology, paleontology, and the rest; used in conjunction with precise physics and geological dating methods.

People can still believe in gods if they want to. But you can't argue with the facts; not if you want to be thought of as educated, intelligent or informed. This is science, and science accepts the Theory of Evolution, expressed in its present form by Charles Darwin in the late nineteenth century, and confirmed, clarified, and updated by experts in that theory and related fields. Darwin was undoubtedly a smart fellow; he even deduced correctly, without good evidence, that man had evolved in Africa.

QUESTIONS EVEN A DARWINIAN CAN ASK

I myself am a Darwinian—to a point. The fossil record, when it is factual, cannot be ignored. Every farmer and animal breeder knows about variations within species. These things are factual. Also, as I said, the theory is logical. But being the skeptical type I still have certain problems with Darwinism— certain questions.

1. The first and foremost of these is the idea of *natural selection by a gradual process of random variation and mutation*. This does not question evolution as such, but the *mechanism* of evolution; but let's save that for a later chapter. Right now, I'll just throw out a few of the larger, more philosophical questions one could ask.
2. *Where did the first life come from?* It is assumed, as stated above, that life arose from the primordial swamps. But is that really certain? The fossil record is there alright, but not convincing proof that it arose spontaneously. Repeating the process experimentally has proven to be elusive. It is far more difficult than was previously supposed—perhaps impossible. The problem is that the further science develops, the more difficult it seems to create complex organic molecular structures, proteins, out of the constituent elements; carbon, nitrogen hydrogen, oxygen, and so on—not to mention the tremendously more complex chains of proteins which make up DNA. There is another stumbling block; the complex structure of living cells and organic systems. But I don't want to steal Michael Behe's thunder.

   If we give credibility to the idea of astrobiology, that life on Earth was seeded from outer space, then we simply transfer the problem to another place and another time but the problem still remains, with one exception; we gain *a lot* more time for evolution to take place. Still the truth of it, the *origin*, remains unknown. You might say that the unknown has always seemed to be "miraculous." Perhaps, but still the question remains.
3. *Where did everything come from?* Not just life, but *everything*—the Earth, our Solar System, the galaxies, the universe? Did it all begin in a sudden

Big Bang as is currently supposed? But how can something come from nothing? Was it always "here?" But how can that be? Where did it come from? Did God create it? But where did God come from?

4. *Where did consciousness come from?* This means that a part of the universe—us—can *think* about the universe, *and can be aware of itself.* A little reflection lets you see how impossible this seems. Materialists (in philosophy), say there is no such thing as consciousness. It is simply a matter of complex molecules. But we all have it—or at least a lot of us have it. What is it, and where did it come from?

The last two questions are perhaps unanswerable questions. Those who believe in God have answers to them, but not scientific ones. In light of these questions, however, those who don't believe in God can't dismiss such beliefs as totally irrelevant to science. But one can not discuss the subject of our origins without thinking of them. In fact, my reason for mentioning this type of questions, which derive from Kant's *Antinomies of Pure Reason* (Kant 1950; Copleston 1994), is to show that at the bottom, there are things that perhaps can never be known. If we can't answer these basic questions, how then can we answer questions about what comes later?

The answer is that we must use science as best we can. We answer those things which we can prove logically using the scientific method and the factual evidence. The more evidence we have, the better. This is what science is—and why science, although limited, is winning out over religion, and even philosophy. You can't argue with the facts!

MICHAEL BEHE UPSETS THE APPLE CART

So what happens when a respected scientist comes along and tries to upset the apple cart? Normally he would simply be ignored by the scientific community and the major scientific journals, or eliminated through the "peer review process." This is where one's academic peers consider whether your work is competent to be considered acceptable for professional journals in your field, and thus, by implication, acceptable to science. But this is not an iron clad system. There are free markets, and there is politics. The peer review system can be bypassed and it can be modified. There is also the Internet. Anything can get on the Internet.

In the case of Michael Behe it worked, I suspect, something like this: In his book on the subject, *Darwin's Black Box: The Biochemical Challenge to Evolution* (Behe 1996), Behe never actually says he is arguing for divine creation when he speaks of *design*. But his supporters and critics alike, it's reasonable to assume, take it for granted that that's precisely what he means. The conservative movement in America today is sizable. A conservative book, if its author is credible, will not have much trouble being published, and has a good chance of a wide readership. To spell it out; many conservatives believe in Biblical Creation. Most conservatives believe in God. *Design* is an idea that appeals to conservatives. It's not unreasonable, then, to say that Behe bypassed the peer review process by appealing to conservatives. When Tom Bethel helps you get published, it's hard to draw any other conclusion.

This is not a bad thing of course. And, as I will argue throughout this book, this, and other ways of bypassing the peer review process may be all that stands between us and the abyss—meaning the demise of science. One reason is politics. Science has become politicized. At this point therefore, I won't say that ulterior motives automatically impact negatively on Behe's credibility. He got published any way he could. But let's see if he's right. Let's look at his argument.

IRREDUCIBLE COMPLEXITY

Behe's theory is that ever since the discoveries of Watson and Crick, the discoverers of DNA, it has become evident that, on the molecular level, biological systems reveal a staggering level of complexity never before anticipated. Under an electron microscope, for example, bacterial flagella look like tiny electric motors replete with stators, rotors, shafts, o-rings, housing, and universal joints (Behe 1996, 71). Cilia are complex machines which include seemingly "ingenious" combinations of mechanical, chemical, and biological principles, in intricate patterns and structures (60). This complexity accomplishes what used to be thought of as a simple process; the hair like extensions—cilia—on some cells, "swimming" in a fluid. Vision also requires a complex biochemical chain or sequence to accomplish what once was viewed—and still is regarded, by many Darwinians—in a purely anatomical sense.

These examples, along with blood clotting and other biological systems, are used by Behe to demonstrate his contention that such complexity could not have arisen by chance; meaning by natural selection. His reason is simple: *Any variation and / or mutation leading up to the present, functioning, biological machine now evident, would not work, and the organism would die.* If so, then gradual, incremental development, of the kind expected by Darwin's theory, is not possible for irreducibly complex biological machines. Some other method is required.

Behe concludes that the method is *design*. On the other hand, Behe has no problem accepting the evolution of species and even evolution on the molecular level—as long as irreducibly complex biological machines are not evident. In other words, some of life evolved and some was designed. It is up to science to differentiate, according to Behe.

CARTOONS AND COMPLEXITY THEORY

Behe is fond of making his point with the aid of well known cartoon, or comic strip characters. One example is the Rube Goldberg machine (RGM) after the comic strip by that name. Typical of the RGM is a complex interconnecting chain or sequence of events and parts, all of which are highly unlikely; culminating in some ridiculous, menial, or funny chore. Behe uses the example of the RGM to illustrate the nature of an irreducibly complex system—a system of numerous, complex steps, all of which are absolutely necessary to the final outcome; for example, blood clotting in mammals.

I find this curious because the defining characteristic of a RGM is *not* its irreducible complexity. Every real machine has a greater claim to that. Take away an essential part of any machine and it will cease to function. It won't work. The thing about a RGM, the thing that makes it *funny*, is the absurdity of expecting such a contraption to work at all. In fact, it *could never work*; except in the world of cartoons.

Real machines, on the other hand, the mechanical kind that engineers build, have a better claim to irreducible complexity. But reality often negates even that claim. Some parts of real machines, such as the "bells and whistles," and safety devices, can easily be eliminated, though we often choose not to do so. Even some critical parts can be done differently or sometimes done away with altogether. Our experience with machines tells us that when we really know how a complex piece of equipment operates, many things are possible— many alternatives. I suspect that the same is true of biological machines. At the same time, one should acknowledge that certain critical parts of any machine, biological or otherwise, can never be done without or removed if the machine is to be viable. This one fact supports the theory of irreducible complexity and may be Behe's lasting contribution to science.

But what of the idea of complexity itself, or *complexity theory* as it is now called? Is there some intrinsic principle in all matter, whether organic or inorganic, which compels it toward greater complexity? This has been a subject of increasing interest in the mainstream academic world in recent years as it has become more and more evident that the traditional Darwinian model cannot be correct. Some theorists have speculated that a "spontaneous organizing principle," instead of random chance, is responsible for the origins of life and its subsequent development. This is theoretically shown to be a natural process which will occur any time certain conditions are met and certain catalysts are present.

On the face of it, this is an intriguing idea. If true, it could explain many riddles and unanswered questions in the accepted Darwinian paradigm. Beside providing an explanation for the impossibly remote chance that the precise mixture of chemicals could somehow have come together by chance to form the first living cells; it could also explain other unanswered questions, such as how, in the Cambrian explosion of 550 million years ago, almost all of the phyla now extant plus many others now extinct, were suddenly formed. While the details of complexity theory are beyond the scope of this discussion, I bring it up to show that there is a wider context for Behe's theory, related questions which were in the academic air when Behe came up with his proposed solution. There may indeed be some organizing principle other than "design" causing irreducible complexity. But my guess is that current versions of complexity theory, such as that of Stuart Kauffman (1995), will not provide it. The theory is still in its pre-scientific or philosophical stages. What is wanted is insight, and more importantly, evidence. Behe provides some. Joseph (see Chapter 14) provides more.

MISSING SCIENTIFIC PAPERS

One of Behe's strongest arguments in support of his thesis is that there are very few scientific papers published which explain how evolution works on a molecular level. Those which do exist are unconvincing. Behe cites the *Journal of Molecular Evolution* for its failure on this account. None of the papers published by the *JME*, he claims, "has ever proposed a detailed model by which a complex biochemical system might have been produced in a gradual, step-by-step Darwinian fashion . . . [N]o one has ever asked in the pages of *JME* such questions as the following: How did the photosynthetic reaction center develop? How did intramolecular transport start? How did cholesterol biosynthesis begin? How did retinal become involved in vision? How did phosphoprotein signaling pathways develop?" The fact that none of these questions have been addressed, or solved, indicates the current lack of understanding of how biochemical systems evolved (Behe 1996, 176).

After searching the professional literature, Behe could find no satisfactory explanations in answer to these, and related questions. His search included such prestigious periodicals as the *Proceedings of the National Academy of Sciences*. A search through authoritative books on the subject proved no more fruitful. Two mentioned were; The *Neutral Theory of Molecular Evolution*, by Motoo Kimura, and *The Origins of Order* by Stuart Kaufman (Behe 1996, 178). I mention these resources so readers can pursue the subject further if they so desire, this being a critical point in Behe's argument. I myself perused the literature and will give an indication of what I found in the concluding section of this chapter.

BLIND WATCHMAKERS

In the nineteenth century, a theologian named Richard Paley made the argument that if someone found a pocket watch on the ground and had never seen one before, it would not be difficult to conclude that this was a man-made object, made by design, as opposed to a naturally occurring object, like a rock, or a sea shell. The same could be concluded about the "designs of God," as for example in that fabulous organ, the human eye. The reasons will not be entered into here, suffice to say that most have been refuted over the years, except perhaps the one which Behe uses—the argument from irreducible complexity.

In the wake of Darwin's famous book, *The Origin of Species*, a bishop named Wilberforce, asked Thomas Huxley, Darwin's famous defender, if he thought he was descended from apes. Huxley replied that he preferred to be the son of a monkey, rather than to be related to anyone as ignorant as the bishop. In America, a few years later, the same type of drama was reenacted in the famous Scopes Monkey Trial. There, Clarence Darrow "made a monkey out of" William Jennings Brian. And so the battle between Evolution and Creationism has continued, over the years, in some circles.

Recently, Richard Dawkins picked up on Paley's analogy, using it for the title of his book, *The Blind Watchmaker*, in which he confidently disposes of the argument from design and all latter day creationists, using the most up to date proofs of Darwin's theory of evolution by natural selection (Dawkins 1996). In turn, Behe disposes of Dawkins, or tries to.

To give an example of the nature of this debate, let's look at the case of the bombardier beetle. When threatened, the beetle squirts a boiling hot liquid at its attacker. The liquid is composed of hydrogen peroxide and hydroquinone, mixed together with enzyme catalysts, to produce a highly irritating chemical called quinone, along with boiling water in the form of a steam jet (Behe 1996, 31). These highly sophisticated chemical reactions, produced at just the right time, in just the right proportions, and without harm to the beetle, could not be the result of evolution, argue creationists. Nonsense, replies Dawkins, and he explains it all very logically, according to Darwinian principles (Dawkins 1996).

But Behe thinks that neither makes their case. To Behe, the key question is this: "How could complex biochemical systems be gradually produced?" He adds, "—the burden of the Darwinians is to answer two questions: First, what exactly *are* the stages of beetle evolution, in all their complex glory? Second, given these stages, how does Darwinism get us from one to the next?" (Behe 1996, 33–4).

Although this is a complex debate not easily decided, one must grant Behe the best of this argument. Neither Dawkins, nor anyone else, perhaps, has ever attempted to explain evolution in the way that Behe demands; that is to say, down to the molecular details. But does this make Behe right? If evolution can't be (or hasn't yet been), explained in this way, does this fact justify the claim that living things were designed?

Tentatively, I think not. The logic just isn't there for it. One may *believe* what one wishes, but science demands proof. If Behe says some living things were designed, he must give us evidence of the designer, or show some design mechanism as Rhawn Joseph does, (see chapter 14), unless he is speaking of *miracles,* in which case he must say so. He can't purport to be talking about science if he really means to speak of religion. This does not, however, subtract from the genuine contribution Behe has made to science in his concept of irreducible complexity.

THE ELEPHANT

In this connection is Behe's idea of "the elephant." The elephant is to Behe, a metaphor for the giant fact or clue staring all of the Darwinian scientists in the face. It is so big, they fail to see it. That fact (the elephant), is to Behe, of course, design. But there is another elephant, or a ghost of one, which Behe himself perhaps fails to notice. Discussing Richard Dickerson's rule for disbarring the supernatural from scientific discourse, Behe asks, where he got such a rule. Is it written in the textbooks? (Behe 1996, 238–41)

Well no; but the principle behind Dickerson's rule lies at the base of all science. Without it we would not be in a scientific age. The meaning is clear. If Behe says there is an elephant here, by which he means design, but more importantly, a designer. He must point out the designer, in tangible, scientific terms. This he does not do.

A CRITIC'S VIEW

Now, let's take a brief look at what Behe's peers have to say about his book. To start with, it is important to point out that mainstream biologists have not ignored him. On logical grounds, Behe's thesis might be said to be the latest, and perhaps the most advanced, form of the old theological Argument from

Design, of which Paley's Watchmaker is the classic example, as was mentioned in above. If Behe's critics truly had no good answer for his arguments, they might have tried a different tactic, of the kind suggested above, (and in chapter one), of the kind used on several of the Principals discussed in later chapters.

One critique of Behe, and *Darwin's Black Box*, is by biologist Keith Robison, and will suffice as an example. Robison attacks the type of analogies Behe uses to illustrate his theory of irreducible complexity, such as the "mousetrap" analogy. (If you remove any component of a simple mousetrap, it won't work.) More importantly, Robison has some specific criticisms of the science. A few will be briefly mentioned.

First, he points out that the human genome contains mostly, apparent "junk." That is, genes, or "pseudogenes," which are useless sequences, with no apparent purpose, which a "designer" presumably wouldn't have included. This argument is weak, because it tries to second guess what the hypothetical "designer" might do, also because it presumes to know, at this early stage of the science, what is, and what is not, "junk" in the genome, but it is telling nevertheless. (One theory is that this "junk" is really "silent genes," a subject which will be discussed in chapter 14).

Another criticism centers on the idea of "cascades," in which one gene or chemical reaction, triggers another—like a chain reaction, or a Rube Goldberg machine. Blood clotting was an example of a cascade, used by Behe. Robison demonstrates that such cascades are not irreducibly complex, either as a whole, or in each individual step along the way. Other species—the example of sharks was given—use modified versions of these cascade chains; and humans also can survive, sometimes without crucial links.

The Krebs cycle, not mentioned by Behe, is a key step in the metabolism of glucose. It is another complex biological system which should be irreducibly complex, but isn't. There are numerous "shortcut pathways," which bypass the complete cycle. Robison asks, if the Krebs cycle is not "irreducibly complex," how can we have any confidence in our ability to recognize such a system when we see one? (Robison 1996).

CONCLUSION

It was not my intention to be overly critical of Michael Behe, who made a valiant last ditch effort to prove that life is the product of design. By this, I believe it is clear that he means supernatural design in the Biblical sense—if not in the Biblical time frame. But this does not mean that his theory of irreducible complexity is wrong. It may indeed prove to be correct.

I said also that I am a Darwinist—to a point. That "point" lets me now say that I too allow the possibility of design—or rather, causality—maybe for philosophical reasons, or for scientific reasons to be discussed in a chapter yet to come. It may well be that nothing happens by accident. That's why I included Behe as one of my Principals. He may yet be right in his essential theory.

Chapter Three

A Little Knowledge:
A Psychiatrist's Crusade

TOXIC PSYCHIATRY

Although people now live longer, healthier lives than ever before, people young and old also still have fears and inner demons. The present chapter deals with the profession which tries to treat these fears and demons, namely, psychiatry. Just as modern medicine has, for the most part, overcome its checkered past, by gaining knowledge of disease and finding cures for most of them, psychiatry's practitioners strived to do likewise by the advancement of "miracle drugs" which have come on the market in recent years. Have they succeeded? The answer to that question is the subject of this chapter.

The Principal of this chapter is Peter Breggin, M.D. In his book, *Toxic Psychiatry*, and many subsequent writings, he makes the case that in the field of psychiatry, the revolution in medicine may be working in reverse; the new miracle drugs and other treatments may harm more people than they help. As might be expected, most mainstream psychiatrists disagree with Breggin's assessment of the value of psychiatric drugs, and place much less emphasis on their harmful effects. They also disagree with him on the efficacy of psychological and social interventions. In recent years, they have come to believe that most or all behavioral disorders are biological in nature and should be treated with drugs. While many psychologists, psychoanalysts, psychotherapists, and other health care workers, still believe that environmental factors can contribute to behavioral problems, they too have come to accept the theories and the drug interventions of psychiatry.

A WARNING

I feel compelled to now issue a caveat and a warning. It is the same type of warning that Dr. Breggin himself issues in the forward to *Toxic Psychiatry*. I do so because I will be discussing theories, based on Breggin's and other experts' research, which could have a profound impact on readers, should they choose to follow suggestions, or advice, which they perceive to be contained, or implied, in these pages. The gist of his warning, which I now pass on to the reader, is this: *Although psychiatric drugs may be harmful or dangerous, stopping them abruptly could also be dangerous, since many of them are addictive, or foster user dependency in other ways. Discontinuing such drugs should only be done under the supervision of an experienced professional.* This warning applies to any Apocryphal medical theory discussed in this book.

MEDS

This chapter is about one expert's claim that psychiatric drugs, and some other related treatments, are harmful to the very patients they purport to treat. I am fully aware that this is an emotionally charged issue, as are other topics discussed in this book. Many people feel they *need* these potent drugs, these legally prescribed medications, or "meds," as they are so cavalierly dubbed in the profession, in order to lead normal lives, to keep the daily stresses and strains of life under control. These are not mentally ill people in the usual meaning of the term; just people who need some help in dealing with the pressures of life. There is "help" for them, and more is coming everyday, provided by an ever growing pharmaceutical industry, and the psychiatric profession. In recent years, that profession and industry have even had the chutzpa to hawk their potions on the national airwaves.

If I can't convince the reader that there is any problem with what was just said—medications to help otherwise normal people "get through the day,"—I can't expect to persuade anyone that "real psychotics" (severely mentally ill people) shouldn't be drugged, especially since this is done ostensibly because "they are a danger to themselves and others."

But let's let Dr. Breggin make his case.

THREE MISCONCEPTIONS

There are three basic reasons normally given for why it is acceptable for society at large, legally or morally to sanction the use of potent drugs for men, women, and children of all ages, for any and all types of mental dysfunction: Reason one is that they seem to help, or to put it another way, the good they do outweighs any possible dangers. Reason two is that they do no harm, or at the worst, the "side affects" are minimal when carefully monitored by a physician. Reason three is that since medical science has proven the various psychiatric diseases or "disorders" to be biological in origin, then it is only proper that these disorders be treated and corrected, chemically, with drugs — psychoactive drugs, to be specific. The bone of contention here, and the essential argument of Breggin, is that these three reasons are misconceptions which are not supported by the evidence.

The ancient belief, from which many present beliefs stem, began centuries, perhaps millennia, ago with the idea that there is something within us, something beyond our control, beyond our own free will, which makes us do evil, which "possesses" us, or makes us "mad" or "insane." It had many names; "evil spirits," "demons," "the devil," "Satan."

Although that view persists to this day, in recent times, the influence of the Age of Science has prompted a search for more rational explanations, for a physical or biological cause for this "evil." What has remained the same, however, is that the evil is still thought to be beyond our control. The overwhelmingly prevalent view of psychiatry today is that psychiatric or mental disorders are biological in origin and not behavioral; hence they must be treated accordingly — with drugs. This is also the conventional wisdom in society, advocated by school nurses, news analysts, and paperback novelists.

In the past, demons were controlled in many ways: the "possessed" were burned at the stake, thrown into dungeons, or put in straight jackets — or, if possible, a holy man exorcised them. In more modern times, and incredibly, still today, these "possessed" were *lobotomized* — that is, the part of their brain which thinks and feels was cut out, removed surgically, or otherwise put out of commission. They are now also drugged, and given high voltage electro-shock.

SCHIZOPHRENIA

The problem with this view, and this course of action, is that it has never been proven scientifically. It is a case of a little knowledge being very dangerous, for that is all psychiatry has when it comes to the human mind — a little knowledge — and acting on that dearth of knowledge, it does a lot of damage. This is one of the great tragedies of our time. I'll make my point, or more precisely, Breggin's point, by reviewing some of the evidence he presents. For the reader who wishes to pursue the matter in greater detail, see Dr. Breggin's many books on the subject—and his website. We begin with *schizophrenia*, the modern name for madness.

It is now widely accepted that schizophrenia is a "brain disease, like multiple sclerosis or Alzheimer's disease" (Breggin 1991, 23). Aside from the long history mentioned above, which led to this commonplace belief, the belief became "scientific fact" to psychiatry and most of the educated world just two decades ago. Around that time for example, active children, especially boys, began to be diagnosed with conditions like "Attention Deficit Disorder," and put on drugs like Ritalin. I bought Breggin's books to find out what was going on.

Popular books promoting the new anti-psychotherapy, pro-biopsychiatry approach, which came out around that time and had great influence over millions of health care professionals, included the following: *The Broken Brain: The Biological Revolution in Psychiatry*, by Nancy Andreasen, in which the author asserts that the major psychiatric illnesses are diseases just like diabetes, heart disease, and cancer. Key to Andreasen's approach and many others of her ilk, is that psychiatrists need not waste time any longer, listening to the patient's problems, but should instead treat the matter as would a neurologist, by asking methodical questions and taking brain scans, and then prescribing a drug. In *The New Psychiatry*, by Jerrold Maximal, the author states that psychiatrists no longer need concern themselves with the theories of Freud and Jung, or with other psychological theories. The public, he writes, has an outdated view of what psychiatrists do (Breggin 1991, 13).

So did I, but I was learning. When my nephew was 17 and under my guardianship, he was diagnosed as a "mild schizophrenic." Since he was not considered a "danger to himself or others," and since the insurance money had run out for the expensive psychiatric institution he had been in, he was released to my care. Although I soon began to apply my own methods, (this nephew is now a college graduate, and a Marine), I also had to follow the edicts of the State as a pre-condition for his release into my care. One condition was that he see a psychiatrist regularly.

Not surprisingly, the psychiatrist prescribed an "anti-psychotic," a type of medication which will be discussed below. Fortunately for my nephew, we didn't continue using this doctor's services any longer than was required.

Breggin states that "while organized psychiatry promotes schizophrenia as a genetic and biological entity, actual research increasingly substantiates its environmental origins" (Breggin 1991, 36). Schizophrenia is a form of psychological collapse, but it's not an "illness," in the usual sense. It is not genetically caused; neither is it caused by microorganisms. Breggin characterizes it as resulting from "schizophrenic overwhelm," or, "psycospiritual crises" (21).

It usually begins in the teen years. To the person in crises, philosophy and theology (though it may not be called that), are matters of life and death. Questions such as, "what's wrong with me," "why am I in the world," "how can I save the situation—or the world," become matters of crucial importance (Breggin 1991, 29). If things get bad enough, if no one cares enough to listen, if those in authority, (usually parents,) instead of helping, add insult to injury, the person breaks down, or "snaps." He or she subsequently says things that "make no sense." He or she "acts out" in an irresponsible, irrational, or a violent way. The young person has become a schizophrenic.

Such a person, in the past as well as the present, if the situation was bad enough, might indeed be "put away," or institutionalized. But as likely, the youth would "get over the phase," or someone would help—a parent, an older sibling, a minister, a friend, a coach—and the crises would soon subside.

Today, many people are still good and caring—as they have always been. But medical science has other plans for the person in crises. He or she is labeled with a "disorder" or a "disease" such as schizophrenia, and given drugs. If he or she resists, it can be imposed on the young person by force, with the full backing of society, through the legal system. And unless he or she is very lucky, this is the beginning of the end for that person.

THE NEUROLEPTICS

These are the drugs inflicted on those diagnosed with schizophrenia, or "mild schizophrenia." They are the "major tranquilizers" or "anti-psychotics," but the preferred name today is neuroleptics. Readers may be familiar with well known neuroleptic drugs such as Haldol, and Thorazine. Neuroleptics are the most frequently administered drugs in mental institutions (Breggin 1991, 51). On many wards they are given to nearly 100% of the patients, and in nursing homes, to around half of the inmates (57). Additional millions (like my nephew), are administered these drugs outside the mental institutions, in clinics, schools, as outpatients, and by private practitioners. They are also administered by social workers to the homeless. Textbooks of psychiatry claim that neuroleptics have an "anti-psychotic" effect, especially on the classical symptoms such as hallucinations, delusions, and incoherence. But they rarely mention the lobotomy-like effects of the drugs. The general apathy and listlessness of the typical psychiatric ward gives a good clue of this (53).

As was mentioned above, lobotomy surgically knocks out of commission the higher functions of the brain, located in the frontal lobes. Neuroleptics have a similar effect; therefore, Breggin calls their use, "chemical lobotomy" (Breggin 1991, 54). "While neuroleptics are toxic to most brain functions," writes Breggin, "they have an especially well-documented impact on the dopamine neurotransmitter system." These are the major pathways from the deeper brain to the frontal lobe. This is the lobotomy effect (56). He adds, "Chemical lobotomy can have no specifically beneficial effect on any particular human problem or human being" (57). And, "only in psychiatry does the physician actually damage the individual's brain in order to facilitate control over him" (59).

This is the dirty little secret of psychiatry. Such drugs are great for controlling the subject. This is why they are used to tranquilize vicious dogs and to quiet the wards in psychiatric institutions and the nursing homes, and why they were used to silence dissidents in the former Soviet Union. But to cure schizophrenia? I'll let the reader decide.

THE DANGERS OF DEPRESSION

Everyone gets depressed sometime. But now it is an official disease of psychiatry—called "clinical depression," and thus treatable by toxic drugs. Try as you will to convince friends or loved ones that they would be better off without a given "anti-depressant," and they will tell you they "need it," and to "mind your own business." There may indeed be more severe and less severe forms of depression, but Breggin's concerns are nevertheless valid.

In the following excerpt Dr. Breggin tells the story of a young college student who was having a little trouble adjusting:

> Roberta . . . was getting good grades . . . when she first became depressed and sought psychiatric help at the recommendation of her university health service. She was eighteen . . . bright and well motivated. . . . She was going through a sophomore year identity crises. . . . She would have thrived with a sensitive therapist who had an awareness of women's issues.
>
> Instead of moral support and insight, she was given Haldol. Over the next four years, six different physicians watched her deteriorate neurologically without warning her or her family about tardive dyskinesia and without making the diagnosis, even when she was overtly twitching her arms and legs. Instead they switched her from one neuroleptic to another. . . . Eventually a . . . psychologist . . . [sent] her to a general physician, who made the diagnosis. By then she was permanently and physically disabled, with a loss of 30 percent of her IQ (Breggin 1991, 70).

"Tardive dyskinesia," along with neurological disorders which are similar to a brain fever called "lethargic encephalitis," and other adverse side effects caused by toxic drugs, were known of from the early days of neuroleptic use.

But they are still downplayed in the textbooks and by the American Psychiatric Association (Breggin 1991, 74), and the public hears almost nothing about them. Countless numbers of hapless people, young and old, are blithely prescribed these "meds" without being informed of the dangers.

Contrary to the case cited above, neuroleptics are no longer generally used to treat depression—possibly because psychiatry is gradually being influenced by Breggin's work—although they are still used in psychotic depression. A variety of anti-depressants have been used over the years; some familiar names are, Tofranil, Elavil, and Adapin (Breggin 1991, 152). Older anti-depressants, such as lithium, are equally toxic—in spite of early advocates of this drug envisioning putting it into the water supply, like fluoride! (173–75). A category of people who are especially susceptible to depression is elderly women. Psychiatry has a favorite treatment for them—equally harmful in its "side effects," most particularly, severe loss of memory. That treatment is electric shock (184–215).

I haven't even touched on the treatment, and damage to, children. A favorite "med" for children, as mentioned above, is Ritalin, (a stimulant also prescribed occasionally for depression). Ritilin is known to cause brain atrophy or shrinkage (Breggin 1991, 308). A newer anti-depressant, Prozac, is used by innumerable, otherwise normal people, to help them "get through the day," to make them feel "happier." Well so do a lot of illegal, "recreational drugs," with no less damage. We will discuss the class of drugs to which Prozac belongs, below.

For the reader interested in finding out more, check out Dr. Breggin's other books, some of which are listed in the References section of this book, or his website, or contact the International Center for the Study of Psychiatry and Psychology; www.ICSPP.org. Also the reader may be interested in examining the work of another writer, the famous iconoclast psychiatrist, Thomas Szasz, who many years earlier challenged the historical and conceptual foundations, and the moral and legal implications of psychiatry. A sample of Szasz's writings is listed in the References section (Szasz 1972).

CRITICS

One of Murphy's Laws should be, if a theory is wrong, or at least vulnerable, it will be widely challenged. It will have lots of critics, and their critiques will be reasonable and well thought out. This was certainly the case with Michael Behe. But if a theory is irrefutable, or at least generally correct, no matter how radical or Apocryphal it may be, its critics will be silent, they will pretend that the theory and its advocates don't exist. Those few who do try to refute a valid

theory backed up by good evidence will necessarily misrepresent it with fee-
ble, specious arguments, and their case will be weak. Here is an example:

In a CABF editorial, (Newsletter of the Child & Adolescent Bipolar Foundation), Martha Hellander writes that she wishes Peter Breggin were right. It
would be easier to finger the drug companies and the APA, than to face the
reality of genetic biological vulnerability to bipolar disorder, (formally
known as manic-depressive disorder). According to Hellander, despite Breggin's "conspiracy theories" claiming that psychiatric disorders are inventions
for the purpose of making huge drug profits, the scientific facts prove otherwise (Hellander 2000).

In the first place, Breggin never claims that psychiatric diseases are inventions. *His argument is that their causes are not biological, and that the evidence cited in making such claims is specious.* Hellander cites a study which
shows brain atrophy in children with bipolar disorder. But *Breggin's argument throughout, is that such clinical evidence, which also includes abnormal
brain scans, is the result of damage done by toxic drugs previously administered.* This is a clear empirical discrepancy, easily resolved by proper scientific studies. But they aren't being done. The original bipolar research, the famous "Amish study," supposedly proving a genetic role in depression, was
shown to be a dubious study, with non-repeatable and non-verifiable results,
proving nothing (Breggin 1991, 145–46).

SSRIS, SUICIDE VIOLENCE, AND MANIA

We will conclude this chapter by mentioning briefly a recent paper published
by Breggin in the International Journal of Risk and Safety in Medicine,
(Breggin 2003/2004). The dominant biological theory of depression (in a nutshell) is that it is caused by a depleted or insufficient availability of the neurotransmitter, serotonin (Breggin 1991, 141).When anti-depressants were discovered to inadvertently block the removal of serotonin from the synapses of
the neurons in the brain, it was concluded that this must be a major factor in
causing their stimulant effect. One result of this discovery was research on
and the development of a drug specifically designed to block the removal of
serotonin. Hence the class of selective serotonin reuptake inhibitors (SSRIs)
was developed; of which Prozac was one of the first, and the best known.

It was also well known that an adverse (side) effect of anti-depressants was
mania (excessive activity, excitable, nervous behavior). Therefore it was
monitored to safeguard the patient against extreme, or psychotic, manifesta-
tions of this effect. With the advent of Prozac and the class of SSRIs, the
known adverse effects grew exponentially. Mania, suicidal thoughts and im-

pulses, violence, excessive nervousness, insomnia, irritability, the inability to stop moving (akathisia), and other aggressive or destructive symptoms began to be recorded in numerous trials and case studies (Breggin 2003/2004). Breggin's paper cites many examples of such adverse effects of SSRIs.

He summarizes the related syndromes in the following excerpt:

1. A relatively sudden onset and rapid escalation of the compulsive aggression against self and/or others.
2. (Obsessive suicidality, violence, etc.) . . . after initial exposure to the medication, or recent change in the dose . . .
3. The presence of other adverse drug reactions . . .
4. Resolution of the syndrome after termination of the causative medication . . .
5. An extreme violent and or bizarre quality to the thoughts and actions. . . .
[6]. An out-of-character quality for the individual as determined by the individual's history (Breggin 2003/2004, 36–37, parenthesis and brackets added).

This is psychiatry. Let the public beware.

Chapter Four

An Old Tactic: A Climatologist Reports on Global Warming

THE USUAL SUSPECTS

This book is primarily about science and not politics or ideology. Yet it would be difficult if not impossible, to discuss the present topic which is global warming, without mentioning the historical context, the motivations, and tactics of certain "usual suspects" who are perpetrating the Big Lie in its present form. As the reader may have already surmised, and will see in future chapters, there are often baser motives at work other than the quest for "pure knowledge," or "disinterested science for its own sake."

In this chapter, politics plays such an important role that I thought it appropriate to preface the science with a brief overview of the ideology and historical backdrop which motivates it. The reader will have no trouble seeing where I stand on the matter politically. Nevertheless it is the science of the issue, and not the politics, which will be the final arbiter.

In the early years of the Communist era, first in Russia and later in China, and then elsewhere, the Big Lie was put to good use. Countries where no individual freedom existed were called "Democratic Republics," or "People's Republics." Nations held together only by Russian tanks and military oppression were said to be part of a great "Soviet Union." Elections were held which offered no choice at all, and no opposing political parties. Collapsing economies were called "prosperous." Misery was called "happiness." Permanent, chronic shortages and hunger were called "five year plans." Mass starvation of millions was called "agrarian reform" (Courtois, S., et. al, 1999).

At long last, this "dialectical" system of lies, contradictions, and death—this "scientific system"—began to implode of its own innate defects, as some economists had said it must (Von Mises 1963). It collapsed from its inability

28

to promote the general welfare of people and society. People began to see the lie for what it was. They began to see that the real scientific political system was one of individual freedom, free markets and democracy—and they began to see light at the end of the tunnel. They began to have hope again.

But where have all the "idealists and progressives" gone, who had carried the banner of socialism for so long? Where were all of the "fellow travelers," who here at home had turned a blind eye to Stalin's and Mao's atrocities, and who in those countries had carried out these "noble" experiments which had, in the names of Marx's and Lenin's "scientific theories," brought about all this ruin? Some, or at least their spiritual heirs, became professors in the universities. Some became government bureaucrats. Some peopled the "Free Speech movement" of the 1960's (Horowitz 1998). The Old World, to be sure, retained its share, and still abounds with legions of these idealists and progressives. In recent years, they have been busy creating a European Union, and forming dictatorships in the Third World, but these activities do not concern us at present.

A few became scientists of a sort, scientists who work within the government and attempt to make policy which would control our lives. Many became environmentalists. It was as if they recognized intuitively that they had a winner here—a new "scientific theory" which the poor, ignorant, common person would not be able to challenge. If communism had failed, environmentalism would not. They took to saying again that it was not about politics, not a matter of opinion—this was science.

PEOPLE WILL ALWAYS TALK ABOUT THE WEATHER

The Principal of this chapter, a climatologist, is the principal and most credible critic of the "science" of these progressives of environmentalism. His name is Patrick Michaels. His first important book on the subject is, *Sound and Fury / the Science and Politics of Global Warming* (Michaels 1992). The reader who has an open mind, who is able to draw conclusions on the basis of the evidence, will have no trouble seeing that he is right. But, as I said, without knowing something about the politics of the issue, and the motivations of the people involved, it would be difficult to see the reason for the controversy in the first place.

Dr. Michaels, like many of us, is an admirer of Thomas Jefferson. Early on in his book, *Sound and Fury*, he cites Jefferson's famous tract, *Notes on the State of Virginia*, as the first example in writing of someone attributing, albeit wrongly, a change in climate due to human industrial or economic activity (Michaels 1992, 1–4).

But I saw something else in this reference to Jefferson. I saw a natural inclination in people, especially older people, (to whom Jefferson referred), to find uniqueness in the weather they have experienced during their lifetimes. It goes something like this: "I never did see such a cold (or mild) winter." or, "I never did see so little (or so much) rain (or snow) in these parts." This is of course an example of *anecdotal* evidence, and it is often meaningless from a scientific point of view. Aside from the fact that the statements may or may not be true or accurate, this type of evidence takes personal or narrow examples, and turns them into generalizations which we are supposed to accept as scientific fact.

On a fundamental level, today's environmentalists are doing precisely that. Just prior to the beginning of the current belief in global warming in 1984, these same environmentalists, were predicting—again on anecdotal evidence —a coming ice age! And then we had one hot summer in the American Midwest and *voila*, the progressives and idealists saw a new winner—a new way to get legislation, to get control. The ice age was out! Global warming was in!

But I've already given my opinion as to the motivation of these environmentalists. Let's look at some of the facts.

FORECASTING THE FUTURE

In 1988, a leading NASA scientist named James Hansen testified before Congress that global warming could be ascribed "with a high degree of confidence" to the greenhouse effect. Within a few days, the Media had revealed the "popular vision" of global warming to the nation. An enhanced greenhouse effect caused by toxic gases (primarily CO_2, but also other gases), emitted by man—especially through United States industrial activity and wastefulness, would cause a rapid increase in the planet's temperature of 7°F. The polar ice caps would begin melting, causing sea levels to rise 25 feet, which would in turn flood most coastal cities. Temperatures in the "corn belt" of the American Midwest would regularly exceed 100°F. Massive worldwide deforestation would occur, creating great new deserts where they didn't exist before (Michaels 1992, 6). Add to this, overpopulation, and the inevitable war over what's left, and you have your usual apocalypse which only your friendly "idealists and progressives" can prevent—if only we agree to give them total control. If you think I'm exaggerating, read on.

As an indication of how this popular vision seeped down into the general population, within one year of Hansen's pronouncements, my son Tom was doing a school science project on this very subject. He was making a papier-mâché model of the catastrophe just described. When I told him that it might not be true, he became very upset. The teacher had said it was so; she wouldn't lie to a twelve year old, would she?

Where did this vision come from? Where did Hansen and his co-workers get their information? They got it from computer models. These models are called General Circulation Models or GCMs. They are useful in helping predict the weather for tomorrow or the next day, or to help students understand the general principles of climatology, but they were never intended for forecasting the future (Michaels 1992, 32). They are far too primitive for that. As an example of the lack of accuracy of GCMs, the first influential computer model simulated temperatures, before greenhouse enhancement, of 9°F colder than present temperatures. That's enough to put Chicago under a glacier (32). John Walsh of the University of Illinois wrote in 1992 that, "currently used GCMs routinely have errors of . . . 9°F to 18°F in the Arctic winter even before increases in CO_2 are considered" (34). But that didn't stop the environmentalists from using GCMs as the bases for advocating the enacting of a myriad of new laws designed to tell the rest of us what we can, or can not, do; how we may live, what we may own, and how we may make a living. In the words of Hansen, "There will be no sudden change. There will be those who don't agree, but as soon as the man on the street notices, it won't matter" (37). The poor schlemiels won't matter.

REAL TEMPERATURES

This section is about what real scientists do. They compare the predictions made by models or hypotheses against real empirical facts. They do experiments. Sometimes, as in the case of the effort to track climate change against greenhouse gases in the atmosphere, the scope of the experiment might span a hundred or even a thousand years or more, in the sense that the applicable data might go back that far. This means that the experiment can't be done quickly. It's a cooperative effort that requires a lot of time and patience. And it certainly means that jumping to hasty conclusions is not good science.

There is no doubt that CO_2 and other greenhouse gases have been increasing over the twentieth century due to industrial activity—primarily the burning of fossil fuels—and will continue to increase. But what will be the effect? Will it be harmful or catastrophic, as environmentalists claim; or will it be beneficial, or somewhere in between? Will there be a large temperature change, or a small one, or none? It takes science to answer such questions, not political agitators masquerading as scientists, not idealists and progressives looking for another chance to grab power.

This is a big subject, so let's start with something small. The NASA record for the past century (1900–1990), shows an average warming of .8°F (8/10ths of one degree Fahrenheit). A similar record for the U.S. from East Anglia University (by climatologists Jones and Wigley), was .81°F, and extended globally, it was around .6°F (6/10 of a degree F)—hardly catastrophic (Michaels 1992, 46). But most of the thermometers from which the temperatures were recorded were in population areas. As the populations grew, the local temperature increased (from body heat, electricity, motors, increased moisture from irrigation, buildings retaining heat, etc.). Land cleared for agriculture also has the effect of raising local temperatures, because the land is bare for half the year and the bare, black soil absorbs more heat from the Sun.

But the average temperature of the Earth may not be rising. One way to find out is to remove the densely populated thermometer recordings from the equation. You still have the vast majority of the Earth's atmospheric temperatures to record, so doing this doesn't "skew" or bias the results; in fact, it does the opposite. Tom Karl, of the U.S. Department of Commerce did such a test. His "Historical Climate Network," recorded temperatures in an area as large as the mainland U.S., and showed *no temperature change in the last 60 years* (Michaels 1992, 45). Add to this the fact that many of the thermometers were put in during the last part of the nineteenth century, at the end of a two hundred year cold spell known as "the little ice age" (43). It's not hard to imagine that temperatures might rise a little from that point. Add also the knowledge that the present weather patterns are part of very long term weather trends of thousands of years, the causes of which are not completely understood, and it is not at all a forgone conclusion that the Earth is warming in any significant way, uniquely cased by modern industrial activity.

An interesting anomaly worth noting comes from Jones and Wigley's record for the Southern Hemisphere (Michaels 1992, 51). They note that most of the enhancement of greenhouse gases came *after* WWII, but most of the modest temperature increase recorded came *before the gases were even emitted.* The Northern Hemisphere, which contains most of the land, and where most of the industrial activity occurred, should have warmed more and faster—but it didn't. In the past 50 years, it warmed 1/4°F, at a conservative estimate, but .9°F between 1915 and 1930, again, *before* most of the extra greenhouse gases were emitted (55).

A LITTLE CLIMATOLOGY

After eight more years of exhaustive research, and in collaboration with another like-minded scientist, Pat Michaels, with Robert Balling, wrote another book, *The Satanic Gases / Clearing the Air about Global Warming.* In it, he continues the line of reasoning he began in *Sound and Fury*, but in a more comprehensive way, and with the benefit of eight more years of data.

Michaels is a professor of climatology at the University of Virginia. Accordingly, he gives us here a somewhat detailed lesson in climatology. At first I wondered why. Then I realized that a person can't really have an informed opinion about global warming—one way or another—without some understanding of the fundamentals of climatology. One can always have an opinion on any subject of course, without benefit of knowledge, as many people do, but it won't be an informed one.

"The Sun shines upon the Earth," writes Michaels, "and the Earth's components, including its oceans and its atmosphere, interact with solar radiation. That interaction is called 'climate'" (Michaels and Balling 2000, 23). Visible, and some ultraviolet light, hit the atmosphere and the Earth, and warm it. A lot of the radiation and heat is reflected away, and would dissipate into the cold of space if it were not for the Earth's atmosphere, which captures much of it, warming the Earth further. This is the *greenhouse effect*. It is a natural and necessary occurrence—a good thing. Without it, the Earth would be a very cold place. So any supposed global warming, above what is normal and natural, is not caused by the greenhouse effect, but by an *enhanced* greenhouse effect. Let's be very clear about that.

Most of the natural greenhouse effect comes from *water vapor*, which captures the heat as it is reflected toward outer space. So water vapor is the most common greenhouse gas. Next is CO_2, which comes from human and animal breathing, forest fires, and volcanoes, and of course from the burning of fossil fuels. Next comes *methane* or "natural gas," which is from cow flatulence and leaky gas lines, among other places. Then comes the *CFCs*, or refrigerants, and then a few other gases in smaller amounts. These are the "satanic gases" of today's environmentalism.

It is reasonable to assume, and Michaels agrees, that *effective* CO_2 levels, that is CO_2 combined with other gasses, have reached 60% of doubling pre-industrial (or post ice-age), levels, and are continuing to increase, although at a slower rate than previously. According to computer models, (the GCMs), this should cause a dramatic increase in the temperature of the lower atmosphere. But "it has not changed a lick in more that two decades—a time when the greenhouse gases were growing most rapidly" (Michaels and Balling 2000, 36). Still, on an intuitive or emotional level you might think that something has to be wrong if the CO_2 levels have almost doubled! This is where science comes in handy. C.I. Mora, writing in the prestigious professional journal *Science*, reports that present levels are only *one-tenth* of what they were three hundred million years ago, that is, when the dinosaurs lived, presumably because there were a lot of volcanoes then, and dinosaurs gave off a lot more CO_2 (56). So CO_2 levels are rising, but not to unprecedented levels, not by a long shot.

Next Michaels debunks claims that the intensity and frequency of tropical storms is increasing due to global warming. And then there is Al Gore's belief that *El Niño* is something new and is caused by global warming. But *El Niño* can be traced back in corals a thousand years, and has nothing to do with global warming (Michaels and Balling 2000, 47). Dr. Michaels' conclusions are backed up by an "ocean of data" found in later chapters, which are beyond the scope of this brief summary.

GREEN AND GREENER

Like the movie comedy "Dumb and Dumber," the "greens" (environmentalist extremists) would be funny if they weren't so ill-informed—and hence dangerous. The irony is that if the policies of the "greens" were made the law of the land, the world would become far *less* green. The socio-economic reason is that deindustrialization (which they often advocate) would make people poor, and poor people would cut down many more trees for firewood, just as they would kill many more wild animals for food. They would also turn on one another much more frequently—hence more wars.

But on the other hand one can understand the concerns of the "greens," for what will happen when the CO_2 levels double over the coming fifty to one hundred years? All indications are that the Earth will become greener. Recent studies suggest that the prevalent notion that plant physiology won't be able to adapt in time to the climate change, thus causing planetary ecocide, is simply wrong. The opposite effect, namely a greening of the Earth, is occurring (Michaels and Balling 2000, 181–83).

AN OLD TACTIC

As with others of the scientific subjects addressed by the Principals of our story, logic and good science will probably not be enough to convince the global warming environmentalists. It seems obvious that this is not as much about science as it is about a way to get control, to get political power to rule over you and me—and of course to continue feeding off the federal gravy train.

A small sample of why I think so—in addition to what has already been said—follows, (from a 1995 IPCC report on floods and droughts): "Warmer temperatures will lead to . . . prospects for more severe droughts and / or floods in some places and less severe droughts and / or floods in others" (Michaels and Balling 2000, 53).

That says it all.

Chapter Five

Fifty Thousand Doctors: Why they Will Never Find a Cure for AIDS

PRECEDENTS

A professor of molecular and cell biology at Berkeley, a member of the prestigious National Academy of Science, and a pioneer in retrovirus research, Peter Duesberg says that AIDS is not caused by HIV and is not even contagious. He claims that long term use of "recreational drugs" is the real cause of AIDS, and what's worse, that AZT, the "antiviral" medication once the most often prescribed treatment for AIDS, has hastened the death of thousands of AIDS patients, and actually *causes* cancer and other AIDS related diseases.

Duesberg has been making these seemingly outrageous claims since 1987. For that he has been reviled and shunned by his peers—his career all but destroyed. But his heretical views are reasonable, based on scientific facts, and compelling. They have convinced many people—and saved many lives.

How, one might ask, could "50 thousand doctors," meaning the whole medical establishment, be wrong? Duesberg cites two factors: 1) the historical over eagerness of the "microbe hunters" to blame every disease on germs; and 2) government control of research which stifles free inquiry and critical thought, and becomes in the process "command science." The reader can readily see familiar patterns in these factors which Duesberg cites to explain the AIDS debacle. In spite of dire predictions, the disease has never broken out of its unusual "risk groups" and spread to the general population. The preventative measures—blood screening, education, condoms, and sterile needles for addicts, once touted, have had no effect. For those inflicted there is no cure, no vaccine, and no ability to predict its course, in spite of billions of tax dollars spent.

Duesberg's book, *Inventing the AIDS Virus*, is the definitive work on the subject (Duesberg 1996). The facts are all there, beginning with a history of the "microbe hunters" who by following the lead and logical theories of Koch and Pasteur (32, 35) had by the early years of the twentieth century conquered most of the infectious diseases known to man. But then many researchers disregarded the objective rules of science set down by Koch to determine if a microbe actually caused a disease. Instead they spent years, at a cost of thousands of lives, blaming microbes for the diseases of malnutrition (37–44).

In Japan, more recently, there was the SMON epidemic in which there was good evidence that an anti-diarrhea drug was causing this polio-like disease. This evidence was suppressed for over ten years by government sponsored "virus hunters" (Duesberg 1996, 11–29). Then, in America, there was the medical empire building and dismal failure dubbed the "War on Cancer," in which many of the current AIDS virus hunters got their start (74–75).

FACTS

Thus there is ample precedent for the current AIDS debacle. But what is needed is proof. Duesberg provides it by the truckload. Chapters crammed with facts, answer the questions some of us had. What was behind Magic Johnson's "miraculous" comeback? What killed Arthur Ashe, Kimberly Bergalis, and Allison Gertz? Was it AIDS or AZT? Following is a sample of the hundreds of facts cited by Duesberg that falsify the HIV hypothesis:

1) Hemophiliacs live *longer* now than prior to the time when three quarters of them became infected with HIV (Duesberg 1996, 286–88).
2) HIV positive *and* negative hemophiliacs suffer the same rates of immunosupression and related diseases (483).
3) Immunosupression in this risk group is acquired by foreign proteins introduced by a clotting factor given to hemophiliacs, and not by HIV (489).
4) AIDS in the U.S. and Europe is highly correlated with risk group categories; e.g., (in the U.S.), 90% male, 98% over 20 years of age, 62% male homosexuals, (almost all of whom used drugs extensively), 32% intravenous drug users (511). This is incompatible with patterns of other sexually transmitted diseases (STDs). For example, syphilis infects both sexes equally and infects teenagers at a much higher rate than does AIDS (518).
5) The risks groups are not compatible with or predicted by the hypothesis of HIV as a sexually transmitted disease, (425), but *are* predicted by the drug hypothesis (519).

6) Natural antiviral antibodies have been the only protection against all other viral diseases since the invention of vaccination in the late eighteenth century. In contrast HIV antibodies are purported by the AIDS establishment to indicate presence of disease instead of the cure, contrary to all medical history and theory, (524), rather than admit that HIV does not cause AIDS.

7) The Drug-AIDS hypothesis is consistent with and predicts AIDS, and many of the 25 plus AIDS diseases which are known to be non-contagious; e.g., Kaposi's sarcoma, dementia, and lymphoma (578). It is also consistent with the explosion of drug consumption among the risk groups in the late 1960's and early 70's. (564). Duesberg's theory, by the way, is that just as tobacco consumption takes 10–20 years to cause cancer, heavy drug consumption causes AIDS by doing irreparable damage to the immune system, in 7–10 years. (412). The medical establishment has not looked favorably on allowing this hypothesis to be experimentally tested, or on releasing the funding necessary (413–14).

8) Several studies indicate that, "Intravenous drug users and male homosexuals (who other studies of the same cohorts show used drugs) lose their T-cells *prior to HIV infection*" (426, emphasis added).

9) Perhaps most unconscionable is the story of AZT. This chemotherapy, a "DNA chain terminator," was shelved years before for its failure as an effective cancer treatment. It was too deadly. It was resurrected by the pharmaceutical company, Burroughs Wellcome, and the National Institutes of Health, (NIH), as a treatment for AIDS. After a rushed, sloppy, and haphazard test series it was approved by the FDA. Subsequent studies cited by Duesberg indicate that AZT is highly toxic, has negative benefits, and some toxic effects are reversible when AZT is discontinued before irreparable damage has occurred. Even more devastating is the following: "Owing to the carcinogenic activity of AZT, the Lymphoma rate of AZT-treated AIDS patients is a staggering 9 percent per year, or 50 percent in three years, according to the National Cancer Institute" (424). In addition to treating patients with full blown AIDS, Duesberg's book documents in numerous case studies that AZT has been prescribed for people, (including infants; see below), *with no symptoms other than that they were HIV positive*—with devastating results. Duesberg presents abundant evidence which demonstrates that the HIV virus is harmless in the absence of high risk behavior, namely the consumption of large quantities of recreational drugs, including heroin, cocaine, nitrites, (poppers), barbiturates, methamphetamines, LSD, and many more, including the medical use of AZT (307–25). As if to prove that HIV = AIDS, the medical establishment encourages the use of these deadly DNA chain termi-

nators, on healthy people who would, and do, live long lives without such therapy (341–48).

10) There are some 18 million HIV positives worldwide. The vast majority, (94 percent), has never had and probably will never have AIDS (425). The number of HIV positives has remained constant, indicating it is not a new virus (542).The number of AIDS cases is increasing, but only in the risk groups, indicating that something else is causing it.

11) Another interesting fact is that, at the Centers for Disease Control, (CDC), there are thousands of recorded cases of HIV *negative* AIDS. It has been renamed, ICL (385–86), but it is merely AIDS under another name. Similarly, an indeterminate number of "ELISA positives," (false positives where a flu virus may react with the HIV test), are included in CDC records of AIDS cases; along with at least 43,000 American AIDS cases never tested for HIV (526).

A POWER GRAB BY GALLO AND FRIENDS

Anyone who has followed this controversy will see what's going on; namely, a power grab and the politicization of medicine. The goal, command science, is "science" dictated by a small powerful elite, whose goals do not always coincide with those of real science.

On the one side: The vast multi-billion dollar tax supported, "peer reviewed," (meaning insider), medical establishment, including the NIH and the CDC; drug companies, AIDS activists, and scientists with conflicts of interest, lobbyist in congress, and royalties on patents for failed medications; insider control of grant monies for research and for which drugs get approved. And, not least important, there is the censorship and control over publication in professional journals crucial to preventing the free exchange of new scientific ideas. Let us not forget, of course, the "willing accomplices in the news media," who presumably want to maintain the status quo with their friends in power, no matter what the cost (Duesberg, 1996). This recurring theme is developed in some of the background chapters of Duesberg's book, and is a major premise of the present work.

Who are the villains? A few notables are; Robert Gallo, Luc Montagnier, Anthony Fauci, Donna Shallala, and "The Pope," David Baltimore (Duesberg 1996; Crewdson 2002. [Crewdson chronicles Gallo's unique culpability]).

On the other side, one man stands out; Peter Duesberg. But he had help from a few brave hearts like Al Regnery, Congressman Gil Gutknecht, who wrote a letter to Anthony Fauci demanding an explanation for the AIDS Hoax, and got double talk from Shallala in reply; and Bryan Ellison, who took on the establishment when Duesberg couldn't get published. There are others; Nobel laureate Kary Mullis, Tony Brown, Robert Root- Bernstein, John Lauritsen, Joan Shenton. The list is growing. They all deserve credit; history may grant them that much (Crewdson 2002; Duesberg 1996; Root-Bernstein 1993; Shenton 1998).

AIDS AFTER TWENTY YEARS

In a benevolent country like ours, where common sense usually prevails and free speech is alive and well, how can it be that the myth that HIV causes AIDS can go on for so long, when it would be so simple to end it? Is it blind fear of a deadly disease which makes people too afraid to see the truth? Is there a conspiracy of silence within the scientific establishment? Or perhaps it can be explained by the greed of monopolies in the pharmaceutical industry acting in collusion with powerful bureaucracies like the National Institute of Health (NIH), the Centers for Disease Control (CDC), and the FDA. Or maybe it's really just the simple practical truth that a terrible mistake was made and nobody has the courage to own up to it.

Whatever the reason, it is clear that this is a case of a travesty, and tragedy, of long duration. But most laymen still know only the Pablum of the official version: that AIDS is caused by HIV, that it is transmitted sexually and through dirty needles, and that you die from it.

Sadly, many health professionals also accept this Pablum without question, though often feeling that there is "something rotten" and that the facts "don't add up." But probably fearing for their jobs, and/or their reputations, they keep silent. Some professionals, knowing something about epidemiology, etiology, and related subjects, strongly suspect that the official version is wrong and still they say nothing. This indeed is sad. It's easy to support these assertions, as we have been doing. But for the gullible, the ill-informed, the easily led lay person, or timid professional, a little backtracking is in order.

HISTORY AND CONTROVERSY

In 1981 Michael Gottlieb examined four young homosexual men dying from an unusual combination of symptoms. These included Kaposi's sarcoma, a rare cancer known since the late nineteenth century; pneumonia; collapse of the immune system; and a variety of opportunistic microbial and parasitic infections. As other similar cases became known, the connection between *gay related immune deficiency* (GRID), as this syndrome was first called (later renamed AIDS), and "poppers," the drug of choice among gays, was noted by James Curran of the CDC, and others. But the idea was soon dropped among official investigators as the less offensive, more politically correct (PC) "virus theory" gained prominence and thereafter became the orthodox view. Since a virus can infect anyone, gay or straight, no one is safe and no particular behavior, is to blame. This PC cliché started a ball rolling which led many people to do the wrong things for the wrong reasons, for a long time, and at the cost of many lives.

The politicization of AIDS research science was early noted by such medical writers as John Lauritsen, Michael Fumento, Tom Bethel, and television commentators Tony Brown in the U.S., and Joan Shenton in the U.K. They pointed out what many medical professionals already knew; that AIDS was not acting like a contagious sexually transmitted disease (STD). Though many differences of opinion would arise among dissenters as to the "real cause" of AIDS, they were in agreement that the official hypothesis was wrong, that it more closely resembled the "party line" of some "controlling authority," (in this case, the NIH and the CDC). STDs *should* infect both genders equally, and all sexually active age groups.

There are risk groups for any disease, of course, but they should reflect the type of risks thought to transmit that disease. AIDS was not fitting, and never did fit these assumptions. For one thing, AIDS never broke out into the general population, meaning the heterosexual community. For another, more than 85% of AIDS victims in the U.S. and Europe continued to be male. Moreover, by the CDC's own admission, it was extremely difficult to transmit through vaginal intercourse. For example, most HIV positive hemophiliacs' wives remained negative.

Scientists soon found mounting evidence that did not fit the "HIV/AIDS equals death," mold. Dr. Joseph Sonnabend, in the early 1980s, saw the clear connection between AIDS and the "fast track" gay lifestyle, which he saw as a mix of multiple sex partners, sometimes thousands, engaged in high risk sexual practices such as anal "fisting," and "rimming," heavy long term use of recreational drugs, including poppers (nitrite inhalants), cocaine, heroin, methamphetamines; in addition to the heavy use of antibiotics prophylactically, to ward off other STDs which were rampant in gay bathhouses and disco back rooms (Duesberg 1996).

Later, professor Robert Root-Bernstein pointed to studies which clearly demonstrated that female prostitutes, with multiple sex partners, many of whom were HIV positive, did not seroconvert (become HIV positive), in the absence of long term use of recreational drugs. He noted that clean needle programs, though he favored them, should not ignore the immunosuppressive effects of the drugs themselves. He also pointed to numerous reasons why

HIV could not cause AIDS in the absence of other "co-factors," such as other microbes or an autoimmune reaction of the body to semen entering the bloodstream directly, as is a risk factor during anal intercourse (Root-Bernstein 1993).

Even Luc Montagnier, the original discoverer of HIV, eventually supported the idea of co-factors, in direct contradiction to the orthodox view. And the famous Albert Sabin, inventor of the Sabin vaccine against polio, became one of the early converts of the most important AIDS dissenter of all—Peter Duesberg. But Sonnabend, Root-Bernstein, and the elderly Sabin, later relented under tremendous political and financial pressure from the NIH, and either moderated their views or quit dissenting altogether. For these scientists and many others, it was either conformity, in effect, obedience, or financial and professional ruin.

In this light, Duesberg is the fountainhead of the dissident AIDS movement. His efforts in defending what he sees as the truth, is nothing short of Herculean. As a member of the very prestigious and mainstream National Academy of Sciences, and one of the leading researchers in the new field of retrovirology, Peter Duesberg could not be easily ignored or stopped by the official wall of silence. Persevering in the face of countless rejections from mainstream scientific journals, he stated emphatically in scientific papers beginning in 1987; that HIV was a harmless virus, (indeed that retroviruses as such were harmless); that the studies clearly indicate that recreational drugs cause AIDS through long term, sustained, heavy use; that poppers cause Kaposi's sarcoma; that heroin, cocaine, Ecstasy, and some sixty other recreational drugs cause cumulative destruction of the T-cells and the immune system; that low T-cell count comes before, not after seroconversion, leaving the body open to opportunistic infections, pneumonias, and cancers, collectively known as AIDS. If that were not enough, he also claimed, as stated above, that AZT, the favored therapy for AIDS, was a deadly "DNA chain terminator," which actually *caused* symptoms indistinguishable from AIDS (Duesberg 1996).

Duesberg, Root-Bernstein, and others have shown that inactive fragments of HIV and antibodies are all that is ever found in AIDS patients. It has never been shown, Duesberg maintains, how inactive fragments of HIV protein found in one out of a thousand T-cells can chemically cause the myriad diseases associated with AIDS. This fact is recognized by the orthodoxy as the "Achilles heel" of the HIV hypothesis. Next, Duesberg refutes the "slow virus," theory which claims that inactive viruses can remain latent and cause disease years later, *without ever reactivating!* In spite of the ten-plus years of "latency" proclaimed by the orthodoxy for HIV to cause AIDS, the FDA approved the use of AZT and other deadly forms of chemotherapy for HIV positives with *no symptoms*! Thus patients who by the CDC's own estimates could stay healthy for ten years (who Duesberg said would remain healthy indefinitely), were given heavy dosages of AZT and died horribly within two years of symptoms which mimicked AIDS and is diagnosed as such. This included many infants! Duesberg, Lauritsen and others, spoke out against these atrocities, with facts and reasoned scientific evidence. Mainstream studies such as the Concord Trial in England and France later confirmed their claims. But AZT continues to be administered to this day, now in combination with so-called protease inhibitors. These drug "cocktails" are so toxic that some members of the AIDS orthodoxy are finally admitting there's a problem. They are calling for "drug holidays" so patients can recuperate, free of drugs (Root-Bernstein 1993; Duesberg 1996; Shenton 1998).

THREE WAYS TO PROVE DUESBERG'S HYPOTHESIS

If prescribed and recreational drugs cause AIDS, one would think that in this era of modern science, it should be quite easy to demonstrate the fact. And so it is. There are three ways to prove a theory or hypothesis, but only one—controlled studies—is considered conclusive. But that way is forbidden by David Baltimore, Anthony Fauci, and other "command scientists" of the NIH, at least insofar as they control the taxpayer supplied grant money to do the research. In spite of the billions of dollars spent to no avail to study the intricacies of dead fragments of HIV, which can only be activated artificially in lab cultures, there is no money to test, in controlled laboratory experiments on animals, the long term effects of heroin and cocaine which are known to be associated with countless deaths, nor nitrite inhalants which are known to be highly toxic to living cells, and to cause Kaposi's sarcoma, nor to test the long term affects of AZT in animals.

In addition to Duesberg's many grant applications, all rejected, Otto Raabe, one of the world's leading toxicologists was repeatedly rejected for grants to study the toxic affects of nitrites on mice. Meanwhile, HIV positive results from questionable "test kits" which are enriching Bob Gallo, Montagnier and a few others, are ruining countless lives, driving some to suicide, precipitating AIDS in previously healthy people through the "prophylactic" use of AZT, only to find out later, as Joan Shenton has demonstrated in her TV specials, that the test kits are no good. Large percentages of HIV positives turn out in subsequent testing to be negative. But the damage is already done (Shenton 1998; Duesberg 1996).

The second way to prove a hypothesis is to examine the available evidence and to create statistical studies of cohorts, risk groups, and populations. This method was good enough to convince the world that tobacco smoking causes cancer, and prompted more controlled experiments to be done. It should be enough to prove that drugs cause AIDS. But resistance by the formidable interest groups here being discussed, and the immense amounts of money involved, dwarfs any power that the tobacco companies once had. A few examples can be given. A study of drug use by homosexuals with AIDS or at risk for AIDS, in five US cities indicates that 79–100% used nitrite inhalants; up to 84% used cocaine, with similar numbers for amphetamines, metaqualone, and several other drugs including alcohol, cigarettes, and AZT (Duesberg 1996, 418). In another study, about 80% of pediatric AIDS cases were born to mothers who were intravenous drug users during pregnancy (419). And a British study showed *no* deaths in a group of 918 HIV positive males who abstained from drug use, and AZT and other "anti-viral" therapies, during a time period (1.25 years), in which the HIV hypothesis predicts 58 deaths (428–29). I suggest that facts such as these, if true, should convince anyone.

Third is the anecdotal evidence. Not scientific, but if one can trust the anecdote to be true, it can also be powerfully convincing. Here too, only a sampling can be related of the many reported over the years.

The most famous, perhaps is Magic Johnson. Symptom free and in great physical condition when first diagnosed with HIV, he began to sink fast with "AIDS" soon after his prescribed AZT regimen began. Then "miraculously" he recovered when he quit taking the drug; enough so to allow him to join the "dream team" in the 1992 Olympics—and win!

Of the thousands of newborn children affected by this tragedy, the case of Lindsey Nagels is heart-rending—and telling. Diagnosed HIV positive soon after her adoption at two months old, she was still a healthy child. Nevertheless her doctor put her on a regimen of Septra, a sulfa drug, and AZT. Immediately, and over the next two years, the child showed increasingly severe "AIDS symptoms," (actually severe side effects of the drugs), including fever, projectile vomiting, wasting away of the muscles, body cramps, skin rash, bone marrow depression. The child was in constant agony. Meanwhile her parents faithfully followed doctor's orders and fed her AZT syrup four times a day. Finally, before it was too late, they heard of Duesberg, and after receiving his help, took the child off AZT. Lindsey's doctor was outraged, but the child recovered and at five years old was in ballet school and doing fine (Duesberg 1996, 306). Many other parents and their children were not so lucky.

In the Dominican Republic, Hector Servino was diagnosed HIV positive after a motorcycle accident. Horrified, his wife committed suicide. Ostracized and vilified, Hector was even refused surgery on his injured leg. Later, however, he was retested and diagnosed HIV negative, but the damage was done (Shenton, 1998).

Another interesting case worth noting is that of Raphael Lombardo who tells his own story in a letter to Duesberg. Found to be HIV positive and discharged by the U.S. Navy, Raphael refused all medication, against the objections of his doctors and his parents. Added to the fear that he would surely die of AIDS was the fact that he was a member of the largest risk group for AIDS—he was a male homosexual who had spent years in the Greenwich Village "gay scene." He had done it all; high risk sexual behavior of all kinds, thousands of partners, and bathhouses. But there was one exception. He never took drugs. Going on stubborn, thick headed, intuition alone, he bucked the system and the overweening authority of everyone around him who insisted that he begin medication immediately. He refused. Ten years later he was still well (Duesberg 1996, 341–48).

Evidence such as this permits me to say that there is no real controversy and no real scientific debate. We are again dealing mainly with power, politics, (which includes corruption and lots of money), and ideology. How does the medical orthodoxy answer these examples and studies, and countless others? With silence. I call that hubris. I also call it an admission of guilt.

"AFRICAN AIDS"

The reader who believes the orthodox view will doubtless still have some objections. The favorite is "African AIDS." Doesn't the fact that in Africa 50% of AIDS victims are women support the official view? Isn't there a "pandemic" in southern Africa as predicted? Duesberg says (Duesberg 2000), there's no pandemic in Africa, where the population is increasing, which would not happen in an epidemic. Traditional African diseases like malaria, TB, malnutrition, amoebic dysentery, and slim disease are now being diagnosed as AIDS because that's where the money is. According to Shenton, HIV tests are not often taken; they're too expensive. Instead of addressing these curable traditional diseases, an AIDS "diagnosis" is made, and is like a death sentence. The victim is ostracized and then loses the will to live. Meanwhile, "condom evangelists" from the CDC lecture Africans on their sexual conduct. A similar story can be told of "Haitian AIDS," a devastating epidemic that never happened (Shenton 1998).

One might ask, if this is all true, why doesn't some independent investigative reporter expose the hoax? The answer is, many have tried, such as those mentioned above, but so far no one in the mainstream media has taken up the challenge; though the opposite is often the case. Dissent has been ignored for twenty years. Mainstream medical writers such as Lawrence Altman of the New York Times, have repeatedly propagated the orthodox view without proof, while turning a deaf ear to the dissident view, never answering, for example, Nobel laureate, Kary Mullis's question: *where is the scientific study which shows HIV to be the cause of AIDS?* Could this be because Altman and other influential medical writers and commentators are not independent? They are members of the clandestine, Epidemic Intelligence Service (EIS), a branch of the CDC, which is said to be the "medical CIA," and which has a vested interest in not being proven wrong in such a big way (Duesberg 1996; Shenton 1998).

HEY O'REILLY!

The truth can't be hidden forever from the American people, or any other people in countries with free speech. Although "medical establishments" are shamefully trying to do just that, they can't succeed. Books, articles, and scientific papers, have been published in the independent press revealing the facts, but the mainstream media, and especially television, where most people get their information, has effectively blocked the dissenting view—the truth—for over twenty years. Eventually, even "pinheads" will see that those who don't take drugs, and never took the "treatment," namely, AZT and the "drug cocktails," *do not die of AIDS*. How long will it take? How long will they be able to keep up the hoax? Who will break the story? Hey, O'Reilly! This Just In! *HIV doesn't cause AIDS, drugs do*. Want a successful war on drugs? Here it is.

Chapter Six

On Classical Ground: Petr Beckmann's Relativity

EINSTEIN AND COMMON SENSE

We pass now from the "mundane to the sublime." We move from practical scientific issues which might potentially affect people's lives, to issues in the next few chapters which at first blush might be of interest mainly to scientists, philosophy students, and historians. But these seemingly divergent issues, as well as others to come, have much in common in that they are some of the best examples of the current crisis in science and its underlying causes.

In this chapter, the principal scientist, Petr Beckmann, attempts to refute Einstein's Theory of Relativity—no small task! But before we get to that, I offer this preamble written on a level of everyday common sense which anyone can understand. As such, it is not a scientific thesis capable of refuting Einstein. Its purpose is to give the reader a better idea of what's at stake and to connect certain logical dots.

The question to be answered is *why* the idea of space-time distortion, so necessary to Einstein's "Theory of Relativity," or more specifically, his *Special Theory of Relativity*, is objectionable philosophically? It's not simply because it is "non intuitive," (as is usually explained), but because it violates the "Law of Identity" or "Non-Contradiction," the logic of Aristotle which made possible two thousand, five hundred years of knowledge acquisition. Without the Law of Identity, no logic and thus no science is possible. But Einstein managed to escape it, primarily through the use of advanced mathematics, and in the realm of objects moving at or near the speed of light. How did he do it? Partly because he was a great genius and convinced a century of physicists and intellectuals that in *this* realm the Law of Identity (implicit in Euclidean geometry and classical physics) didn't apply, or if it did, it was now

raised to a higher plane of sophistication, a level which shielded it from logical scrutiny—and partly because few people understood him.

As we will see shortly, there were notable exceptions to this "century of
physicists" who never accepted the honor of becoming Einsteinians. It nevertheless is my contention that even Einstein couldn't have pulled off such a
stupendous feat without Immanuel Kant first having laid the philosophical
groundwork. It may surprise the modern science enthusiast to be told that
Einstein's theory has philosophical roots, but I believe this is clearly the case.

Before examining some of the evidence which supports these claims, we
should consider first what the essence of this fundamental law is. The law of
Non-Contradiction is the generally understood idea that a thing, or an attribute of a thing, *can not both be (exist) and not be at the same time and in the
same place*. That's what is meant by the Law of Identity or the statement "A
is A." Though this Aristotelian principle is well known and has a very long
tradition, my primary exposure to it was through the works of Ayn Rand
(Rand 1992; Copleston 1993, 283). As you will see, two of the concepts
which form the very fabric of the Law of Identity, "time" and "place," are
called into question by Einstein's "Theory of Relativity."

Let's look at the matter simply and plainly, to get a feel for it, and to see
how it conforms to our general sense of logic. Beckmann's arguments, and a
review of the evidence, will follow.

A IS NON-A

The distance between objects is the measure of the spatial relationship between two fixed points. But planetary objects are not "fixed points." So how
does one measure the distance between them? One way is by using the speed
of light as a "constant" ("c"), against which all else is measured. The logical
requirement is that speed, or velocity, is a measure of *distance divided by
time*—thus even a "constant" velocity would have to yield to this prior requirement. Tom Bethel suggests that since distance or space, and time, become amorphous in the Einstein theory, they can not be the basis of any definition for velocity as used in that theory (Bethel, June 1999).

Einstein could not rely on Kantian subjectivism alone to form the basis of
his revolutionary new scientific theory. He needed concrete, practical reasons. The most significant of those he offered were as follows: a) Preliminary
evidence, which he took to be conclusive, that "c" was indeed constant *between inertial frames*. The Michelson-Morley experiment was paramount in
this regard. b) In view of this evidence; Einstein knew that if light traveled at
a constant speed *between* various inertial frames, the laws of physics would

not apply equally in all inertial frames as the *principle of relativity* requires, unless space and time were *not* constant. c) Lorentz found that elementary particles contracted at high velocities. Einstein interpreted all this as only a Kantian could—*that space and time themselves contracted* (Einstein 1936).

In Einstein's solution, whereby he abandoned objective reality far too casually, but not surprisingly, objects moving at high velocities contract space itself along with the objects moving in space, making the same distance *shorter* for an observer on a rapidly moving object or spaceship, relative to an observer moving slower or at rest. A is non-A.

As *space* contracts, *time* slows down for the rapidly moving observer but this is not noticed except by a more slowly moving observer, or one at rest, and when they later compare "clocks," or evidence of aging. Thus one observer's "time" is different from another's when they rejoin in a common "inertial frame," that is, when they resume moving at the same speed. A is non-A.

According to the Einstein theory, the speed of light is not simply constant as measured between two "fixed points" or between theoretically established coordinates in the orbits of planets, moons, or comets, and observers on Earth. It is "constant" relative to any individual observer in any "inertial frame," (i.e., traveling at whatever speed), on the Earth, the Moon, a rocket ship, a train, an observer hitching a ride on a hypothetical sub-atomic particle, etc.

Thus if observer #1 passes observer #2 at nine times his speed, light continues moving toward, or away from, them *both* at velocity c, even if #1 is moving at 9/10 c, and #2 is moving at 1/10 c. Meanwhile observer #3 in Poughkeepsie, who is at rest or moving uniformly with the Earth, also measures light coming toward him at c, the same constant speed. All three observers then, are all "at rest" from the point of view of their own "inertial frame" as the principle of relativity dictates. They all measure light emanating from the same source, say the Sun, at the *same exact speed relative to each of them regardless of their various rates of motion*—though simple arithmetic should tell them something quite different.

Visualize two trains passing each other in opposite directions each going 50 Mph. An observer in train #1 will feel himself "at rest," and perceive train #2 passing him at 100 Mph. Not so with light. No matter how fast you go in either direction, the same light approaching you, or emanating from you, approaches or emanates at the same constant speed. So if light was the train moving away from you at 50 miles per hour, you could chase it on your Harley at 49 Mph, and it would still continue to recede from you not at 1 Mph as you might expect, but at a rate of 50 Mph. And if it were coming toward you at the same speed, it would gain on you not at 99 Mph, but still at 50 Mph. A is non-A.

Another "relativistic" issue, this one from Einstein's *General Theory of Relativity*, is the equivalence of gravity and inertia. According to Einstein, the two are identical. So my weight of 220 lbs., (or whatever), the pull of my body "mass" toward the center of the Earth, is the same as my weight being forced outward as I am being spun around in a carnival ride called, "the sling." When the sling spins around fast enough, the inertia of centrifugal force pushes me against the outer edges of the ride at a force equaling 220 lbs. The same will happen when I accelerate rapidly or stop short in a vehicle. The "G force" will simulate gravity. The two are equivalent, but are they identical as Einstein claims? Gravity is the pull toward the center of a planet—usually Earth for us, smaller objects have gravity too but it is barely measurable—while inertia is the resistance to acceleration or any change of speed or direction from uniform motion of the inertial frame. A is non-A.

In the Einstein theory, space is "curved." But what does that mean? Any Einsteinian can explain by citing the "bowling ball on the rubber sheet" analogy. The sheet is fastened at the edges, like a trampoline, and the heavy ball resting in its center forms a depression, and *this* is likened to the "curvature of space" in a gravitational field. This is also sometimes referred to as the "fabric" or "geometry" of space (See Chapter 10 for more on this). A is non-A.

Then there is the maxim of "four dimensional space," which has spawned the innumerable other "dimensions" of science fiction. The fourth dimension in the Einstein theory is *time*. Length, width, and depth are the three dimensions of objective reality. The Einsteinians have another "thought picture" for explaining the fourth dimension. Imagine people living on a "two dimensional" world, like images on a sheet of paper or on a movie screen. These "flat people" would find it impossible to visualize the third dimension, depth, because it would not exist for them, it would not be part of their reality. The assumption of course is that *we* can accept the possibility of a world of "two dimensions." Thus by extrapolation, we can also visualize a world of four dimensions.

But the assumption is wrong. This thought picture, as others used by Einsteinians, ignores the principles of causality and logical consistency. We can imagine no such thing to be possible. Images on a sheet of paper, like cartoon characters, can not come to life. But it is imagined that if you looked at such two dimensional "people" sidelong they would disappear. Very Kantian, but it doesn't wash. Remember, all images have reality *as images*. As such they are composed of, or are attributes of, some substance, and are thus part of this three dimensional world. Turn a sheet of paper on edge and it doesn't disappear since it has *some* thickness. A is non-A.

Finally there is the *principle of relativity* from which Einstein named his "Theory of Relativity." This principle, known to Newton, was formulated by Galileo in his *Dialogues*, when he asks what will happen when a ball is dropped from a tall mast on a uniformly moving or still ship. Will the ball drop back, now freed from the ship's momentum, as the Aristotelians of his day claimed, or follow the ship's "inertial frame" and land exactly below at the foot of the mast? (Hall 1981, 50). The fact that the Aristotelians were actually right, in the instance of a moving ship, does not subtract from Galileo's insightful formulation of the principle of relativity. The ball *would* follow the ship's inertia long enough to land fairly precisely at the foot of the mast as Galileo predicted, but, if the mast were tall enough, soon the Earth's gravity, not to mention wind resistance, would prevail and the ball would indeed fall back, how far would depend on how fast the ship was moving, the height of the mast and other factors.

The reason is that the Earth and its gravity is the overriding or *dominant* inertial frame we live in (Hall 1981, 51). But the principle holds nonetheless. It explains the examples mentioned above. It is also why you feel like you are not moving though you may be traveling at a cruising speed of 500 Mph in a jet liner, and the book you drop does not fall back at 500 Mph, but instead lands at your feet. It is also why we can't feel the Earth moving. *Stated simply, as long as an object is moving at a steady rate, it doesn't matter what the relative speed is, it is in an inertial frame and the same laws of physics apply as in every other inertial frame.* This being the case, there is no absolute motion, no "preferred frame," and no absolute state of rest. This then is the principle of relativity.

Note that an object traveling with uniform motion on Earth is only in a *limited* inertial frame. It will cease moving (relative to the Earth) as soon as the Earth's dominant gravitational force takes effect—unless some other force (e.g. electrical, mechanical, hydraulic, etc.), continues to act upon it. The point is that in ignoring this distinction, the importance of the gravitational field was often overlooked, as we shall see. As Beckmann explains it; in all of the experiments used to prove the Einstein theory, the "observer" was always stationary in the Earth's gravitational field, so that the experiments "could not reveal whether the observed effect was associated with observer-referred or field-referred velocities (Beckmann 1987, 6).

Einstein turned the principle of relativity on its head and called it "The Theory of Relativity." In the Einstein theory, the *"observer"* is the absolute inertial frame against which all else is measured, not just the speed of light, and as a consequence the speed of everything else, but also space and time themselves. In quantum physics, the subjective observer's importance is even more paramount, but that topic will be discussed in another chapter. Although the observer can not simply be whimsical since he has to follow Einstein's rules, and although Einstein's motives may have seemed reasonable, I submit that his solution, lauded by modern philosophers and intellectuals who turned him into a veritable saint, is typical of Kantian subjectivism. While the principle of relativity fits into an objective, Euclidean world, the "Theory of Relativity" belongs to a world of floating abstractions and Kantian "phenomena." A is non-A.

KANT AND THE SUBJECTIVITY OF SPACE AND TIME

The philosopher Wilhelm Windelband, in discussing Kant's conception of space and time writes: "universitality and necessity . . . is intelligible only if *space and time are nothing but the necessary Forms of man's sensuous perception*. If they possessed a reality independent of the functions of perception, the *a priori* character of mathematical knowledge would be impossible. Were space and time themselves things or real properties and relations of things, then we could know of them only through experience, and therefore never in a universal and necessary way. This last mode of knowledge is possible only if they are nothing but the Form under which all things in our perception must *appear*" (Windelband 1950).

An analysis of the finer points of Kant's philosophy not being the purpose of this essay, I am here trying to impart only a sense of Kant's theory on this subject, by citing a few of the clearest passages pertaining to it, by Kant himself, or commentators on Kant's thought. In his survey of Kant, Copleston states the matter thus: "inasmuch as space and time are *a priori* forms of human sensibility, the range of their application is extended only to things as appearing to us. There is no reason to suppose that they apply to things—in themselves, apart from their appearance to us. Indeed, they cannot do so, for they are essentially conditions for the possibility of appearances" (Copleston 1994).

And Kant himself proclaims: "But if I venture to go beyond all possible experience with my concepts of space and time, which I cannot refrain from doing if I proclaim them characters inherent in things in themselves . . . then a grave error may arise owing to an illusion, in which I proclaim to be universally valid what is merely a subjective condition of the intuition of things and certain only for all objects of senses . . . namely, for all possible experience . . ." (Kant 1950).

What Kant means, and as I will try to show, Einstein accepts, is that if space and time are objectively real things or attributes of real things, then they are "things-in-themselves," (*noumena*). (Please note that what Kant means by "objectively real things" is the exact opposite of the common sense meaning of the term). Therefore, we can know nothing about them. We can only know the things of "experience," (*phenomena*, or things as they appear to the senses). Thus space and time are not real *because* we experience them. It was an easy step then, for Einstein to turn space and time into elastic, amorphous concepts to suit his theory, because the hard, common sense notion of space and time wasn't real. Kant had proved it!

The point of digging up old bones is that what they represent often goes on haunting us for centuries. Kant's theory of space and time was not about physics but about metaphysics, which like it or not, lays the foundations of science. But if a theory is irrational or ill-conceived in any way, we pay the price. Ironically, Kant's theory may have been much more conservative than Einstein's. Although grounded in subjectivism, Kant had no reason to doubt the eternal verities of a common sense space and time. But when concepts are removed from their referents in reality anything can happen. Well grounded concepts and even axioms become floating abstractions. Pandora's Box, once opened, is hard to close.

EINSTEIN, PHILOSOPHICAL IDEAS

The following section leaves little doubt that Einstein was a Kantian, and that his conception of physics, in particular his conception of "warped" space/time, was strongly influenced by Kant's philosophy.

Einstein believed he had a duty to challenge the philosophical fundamentals of physics because they had reached a conceptual dead-end. He writes that at a "time when the very foundations of physics have become problematic as they are now . . . the physicist can not simply surrender to the philosopher the critical contemplation of the theoretical foundations for he himself knows best" (Einstein 1936, 290). Moreover, he must critically analyze "the nature of everyday thinking," but "the concept of "real external world" of everyday thinking rests exclusively on "sense impressions" which for a Kantian which I believe Einstein to be, are essentially subjective. To him, the concept of "bodily object" is not correspondent with "sense impressions," but is a "free creation of the human mind" (291). "We attribute to this concept of bodily object a significance, which is to a high degree independent of the sense impressions which originally gave rise to it" (291). These mental creations "appear to us stronger and more unalterable than the . . . sense experience itself, the character of which as anything other than . . . illusion or hallucination is never completely guaranteed" (291). The very fact that we can put this unreliable array of impressions, images, and illusions into a comprehensible order is "an eternal mystery" (292). "It is . . . the great contribution of Immanuel Kant that the postulation of a real external world would be senseless without this comprehensibility" (292). Lest they be misunderstood, comprehensibility and order, according to Kant and Einstein, are not *discovered* in nature; they are *invented or postulated* in the human mind out of the chaos of the external world.

"Nothing can be said *a priori* concerning the manner in which concepts are formed and connected, and how we coordinate them with sense experience" (Einstein 1936, 292). But the pragmatist in Einstein tells him nevertheless that "success alone is the determining factor" (292). Of course how we can ever be sure of our "success" when we can't be sure of our "sense experience," is another "eternal mystery." However, to attain comprehensibility "all that is necessary is to fix a set of rules," (292), like "the rules of a game in which the rules themselves are arbitrary, it is their rigidity alone which makes the game possible" (292). Einstein's goal was to replace the rules of Euclidian geometry and classical physics with his own, and so far he has succeeded.

"An important property of sense experience (remembering that these are "free creations of the human mind"), is their temporal order. This kind of order leads to the mental conception of subjective time" (Einstein 1936, 295). And, "The bold notion of 'space'...transformed our mental concept of relations of positions of bodily objects into the notion of the position of bodily objects in 'space'" (297).

"The fatal error" of Euclidean geometry which the Einstein theory replaces is "that logical necessity (precedes) all experience" (Einstein 1936, 298). "The empirical basis on which (this) axiomatic construction of Euclidean geometry rests had fallen into oblivion" (298). "It concerns the totality of laws which must hold for the relative positions of rigid bodies independently of time. As one may see, the physical notion of space also, as originally used in physics, is tied to the existence of rigid bodies" (298). In other words, instead of calculating the relative positions of planetary bodies (for example), by the best and most logical methods then available, making that do until more accurate methods and instruments could be devised, and employing them *as if they were "fixed positions,"* for the purpose of representing reality as accurately as possible, Einstein chose instead to throw out the entire notion of objective time and space and replace them with a subjective Kantian conception.

We are given a choice between the so-called "a priori" absolute of Euclidean space and a subjective invention, forgetting that Euclidian geometry and classical physics are merely tools to give us a clean or simplified, symbolic representation of the way things are and what we see, like a blueprint. We build skyscrapers, jets, and computers, using such concepts. But a subjective invention, no matter how odd it seems, how contrary to logic, no matter how much it violates the law of Non-Contradiction, is instead offered up as if by *leger-demain*, to explain all of the factual evidence, forgetting that to a Kantian, factual evidence doesn't count for much, as we shall see.

There is no need to recount the scientific steps which Einstein took to arrive at his "Theory of Relativity." We will touch upon some of them as they pertain to Beckmann's alternate hypotheses. They are readily available to anyone interested. In fact, they are practically common knowledge, differing levels of understanding notwithstanding. Every child knows $E = mc^2$, and every college graduate knows that Einstein's theories were all "proven," though what the proofs are, is perhaps rather vague. In fact due to the breakdown of basic logic where this subject is concerned, most physics majors and professionals who don't specialize in "Relativity," are frankly indifferent to the subject (Beckmann 1987, 20), as they would be to any subject which defies common sense, such as modern philosophy. Only intellectuals are enthralled, as they are with any irrational idea which allows them to escape reality based limitations. They seem to defend it as they would a personal assault on their honor, and view any attack on Einstein in a hysterical manner (Bethel, October 1993 and June 1999).

One further step Einstein took is of philosophical interest, and then we shall move on. He asserts, "There is no inductive method which could lead to the fundamental concepts of physics. Failure to understand this fact constituted the basic philosophical error of so many investigators of the nineteenth century" (Einstein 1936, 307). Einstein also criticizes J. S. Mill, a proponent of logical induction, in this regard (303). "Logical thinking is necessarily deductive; it is based on hypothetical concepts and axioms." "The most satisfactory situation is to be found in cases where the new fundamental hypotheses are suggested by the world of experience itself."

But isn't this last statement an example of induction, which can never lead to "fundamental concepts of physics"? Isn't Euclidean geometry, and classical Newtonian and Galilean physics, necessarily deductive, at least in that aspect of it of which Einstein is so critical, and based on axioms "which precede all experience?" And isn't experience, or sense impression, the basis for induction and liable to "illusion and hallucination?"

Here, in short, we see the most highly touted intellect of the twentieth century in a state of philosophical confusion. There is a strong leaning toward Kantian subjectivism, especially as concerns the fundamental epistemological concepts. There is a hint of pragmatism, and there is scientific realism. Perhaps Einstein is really just an amateur in philosophy and out of his depth, but I don't think that's it, because the professionals agreed that "he knew best," and kept him on a pedestal for a hundred years. As stated above, there were practical scientific reasons, landmarks, dilemmas in physics, famous experiments, and dead-ends, which prompted Einstein to search for a new and original theory. But how did he employ his genius? With some significant exceptions, he turned to the world of the surreal into the realm of "Dada physics" (Bethel, August 1993).

Eventually, Petr Beckmann, picking up where Lorentz and others had left off, demonstrated that original concepts would come which retained what was valid in classical physics, which was most of it, and corrected it where

needed. These concepts would make use of the latest findings in electrodynamics, and the most sensitive, state of the art instrumentation. In fact without sacrificing objective reality, the Law of Identity, or the principle of relativity, they would satisfy all of the claims made by Einstein's "Theory of
Relativity"—plus two.

A PARADOX

Before going on to Beckmann, I'd like to mention one other item of interest.
Bethel who wrote about Beckmann more than once, calls it, "Dingle's Question," after the physicist Herbert Dingle, one of an impressive list in that field
who didn't accept Einstein's theories. It concerns a paradox or contradiction,
one of the most damning of those resulting from Einsteinian "Relativity" discussed above.

Two observers are traveling at different speeds. According to the principle of
relativity, each is "at rest" in his own inertial frame, with respect to the other,
watching the other travel. In Einstein's theory, the moving observer's clock must
be going slower than the one at rest. But which is which? Note that the *slower*
the clock, the *shorter* the elapsed time for the trip relative to the stationary observer. Thus the stationary observer may be old by the time the traveler moving
near the speed of light returns, but which one is really "stationary?" Which one
should we expect to see old while the other ages only a few days?

So Dingle reasonably asks, "Which clock runs slow?" An Einsteinian glosses
over the problem saying the clocks can only be brought together once, "at the
moment they pass," thus preventing a contradiction (Bethel, April 1999). But
one can easily envision an experiment whereby two spaceships traveling at two
different speeds relative to Earth, compare clocks at the end of their journey, and
check for signs of aging. Since, according to the principle of relativity which
Einstein accepts, the Earth is no privileged frame, even if only one spaceship
travels, Dingle's question is valid. But if atomic clocks do indeed slow down at
high velocities when traveling through denser gravitational fields, or at higher
speeds traveling through the same dominant gravitational field, as Beckmann
explains it, does time slow down nonetheless? The answer is that there are other
laws operating here, but they're not Einstein's.

PETR BECKMANN'S RELATIVITY

Enter now into the world of objective science. It has existed for hundreds of
years. Successful outcomes would be near impossible without it. But due to

the havoc wreaked by modern, Kantian philosophy, even the "hard sciences" can go awry at the theoretical level, not to mention the long standing effect of that philosophy and its spin-offs on the "social sciences," and its more recent connection to "junk science".

This state of affairs was at its height during the first years of the twentieth century when philosophical relativism came into vogue (Johnson 1994). During that period, specifically, in 1905, Einstein wrote his famous paper on Special Relativity which was accepted enthusiastically by intellectuals.

Here then, is objective science at the theoretical level. I remind the reader that *quantitative* proofs, meaning mathematics, are, except for some elementary description, not part of the discussion. For the interested specialist, the source material will provide ample quantitative proof plus leads to more.

1. *Speed of gravity*: Beckmann assumes "with Einstein and practically every other gravity theoretician" that the force of gravity propagates outward from the center of a mass, with the speed of light. (This assumption has now been questioned by some scientists; see Chapter 10). In the electric analogy, confirmed by experience, if we remove a charge, the removal of the electromagnetic force emanating from a fixed point, is propagated at the velocity of light. This is a conventional and measurable assumption (Beckmann 1987, 25). Since it is difficult to "demass" an object large enough to have a measurable gravitational force, the actual speed of gravity hasn't been confirmed by experiment.

2. *Speed of light*: "When light is emitted by a source moving uniformly through a vacuum, its velocity is constant: but with respect to what?" With respect to the observer "regardless of their velocities relative to the source," says Einstein. This is the generally accepted answer in spite of the absence of *direct* proof, (27) and in spite of the fact that it flies in the face of logic, as described above.

3. Beckmann's alternative hypothesis is that "the velocity of light is *constant with respect to the local gravitational field through which it propagates*" (Beckmann 1987, 27). If the Sun were the local rest-frame, light propagating from Earth would first move from Earth at a speed of c, *plus* the speed of Earth orbiting the Sun, (c + v). "Beyond this simple consequence of Galilean relativity, the experimental evidence (bending of light rays in a gravitational field) suggests that the velocity of light varies with the intensity of the gravitational field" (28). "There is also hard . . . evidence that the velocity of light remains constant with respect to the gravitational field but not with respect to the Earth rotating in it" (28).

3. *Some properties of light confirmed by experiment, and the ether considered*: "There is aberration of light as the ray from a star enters the gravitational

field of the Sun," and "a further aberration as the ray passes from the grav-itational field of the Sun to the Earth" (32). The analogy describing aberra-tion is the way vertical rain leaves slanted tracks on the side window of a moving car (30). Thus the moving gravitational fields cause the light ray to slant—very slightly. This is not the same as refraction, or bending, which occurs whenever the density of the medium through which light is traveling, changes. Moreover, it is interesting to note that aberration and refraction are often misinterpreted by Einsteinians as proof of the curvature of space.

Experiments measuring "the coefficient of drag" through a moving medium demonstrate that light moves slower through a medium, and speeds up, relative to the slower speed, when the medium is moving in the direction of the wave. Discovery of the coefficient of drag in the 19th century "caused the scientific community to be unshakably convinced of the physical reality of a partially entrained ether that only cranky maver-icks could doubt" (Beckmann 1987, 35). And, in case the reader is won-dering, Beckmann considers the "entrained ether" as a confusing and un-necessary synonym for, "gravitational field" (37).

4. *Binary stars, moving mirrors and wave theory*: Measurements of light from double stars refute the "ballistic theory" but do not necessarily con-firm the Einstein theory. If light was "particles" traveling like "little bul-lets," its speed from binary stars would be different for each of the two stars because when emitted, one star would be orbiting away from us and the other toward us. (Remember c + v). But light from binary stars is measured at the same speed, seemingly confirming Einstein's theory. Beckmann later modified this view in his newsletter, *Galilean Electro-dynamics*, (Bethel, August 1993). Beckmann's explanation is that light emanating from binary stars initially starts out at different speeds, but soon stabilizes at one common speed as the gravitational fields merge into one, then changes to a different, but common, velocity as it passes other dominant gravitational fields (Beckmann 1987, 37). The fact that light is not "corpuscular" or "ballistic" has been demonstrated by nu-merous "moving mirror" experiments. Light "particles" should bounce off moving mirrors like tennis balls off a racket, and thereby increase their speed (c + v). They don't, because light is a wave.

Waves, whether in water, or sound in air, or light in *its* medium, the gravitational field, have certain things in common. One is that they travel at a certain *constant* speed which is determined by the medium they prop-agate in. A wave made by a moving or stationary object always travels at the same speed *relative to its own medium*. But if we view light from some *other* inertial frame, then its velocity can be added, just as is true of any other wave. Think of a jet catching up with its own sound waves, i.e., breaking the sound barrier. It is not much of a conceptual leap to think of a rocket ship catching up with light waves and thus breaking the light barrier. The method of propulsion to achieve such a feat is of course another question, but it may not be as difficult as is now believed. In the meantime, if we think of light in simple terms of classical waves, I think we can begin to understand how light, and the electromagnetic spectrum, operates in reality.

5. *The famous Michelson-Morley experiment (1881)*: This experiment which was designed to detect the ether—and failed—but it really only refuted the "unentrained ether" theory. This was a hypothetical ether that supposedly permeates all of space like an ocean or an atmosphere, and "ripples" or exhibits "wind resistance" when planets move through it (Beckmann 1987, 39). It did not refute the "entrained ether" theory. The little known *Michelson-Gale* experiment (1925) detected this ether, and it positively demonstrated that light slows down in a gravitational field *relative to the* Earth *rotating in it*. The Earth doesn't make a "big ripple" in the gravitational field because it takes the field with it, though the field does not rotate with the Earth.

This ether or gravitational field can be detected, as was shown by Michelson-Gale, by measuring the difference in light's speed between east-west (faster), and west-east (slower), travel (Beckmann 1987, 44). The Einstein theory also notes the "fringe shift" which Michelson-Gale detected, but attributes it not to a change in light's velocity, but to a change in the fabric of time (45). Bethel brought up this point, citing new evidence, but an Einsteinian replied in a correspondence that light travels *farther* west to east to catch up with the Earth's rotation! (Bethel; August 1993, October 1993 and April 1999). But this correspondent's premise relies on a "privileged frame" which the Einstein theory rejects.

This of course, means throwing out the principle of relativity, demonstrating once again the philosophical dilemma of modern science. Remember that Einstein's theory claims that the speed of light is always constant regardless of the inertial frame of the observer. Had the original Michelson-Morley experiment detected a fringe shift, Einsteinians would probably have made the same argument or excuse. Light, between two fixed points on Earth in the experiment, they might have claimed, has to travel further when the Earth is moving away from the Sun than when it is moving toward the Sun; precisely contrary to Einstein's theory which holds that there are no absolute, "fixed points" in "Euclidean" space, and no privileged frame.

Not surprisingly, Albert Michelson was another great physicist who never accepted Einstein's Theory of Relativity.

6. *On electromagnetic force*: "There is no action at a distance known to us other than the gravitational and electromagnetic and they both obey the same force law" (Beckmann 1987, 46). "The magnetic force between two charges is so small compared to the electric force between them that it is not measurable unless the latter is neutralized" (46). "We are not able to detect a magnetic force unless . . . we remove the electric force that overshadows it," by neutralizing at least one of the two currents (47). "In the macroscopic world we live in . . . the ion grid of the electrically neutral conductor" . . . is . . ."the Earth in our neighborhood. The dominant force field is therefore, the gravitational field . . . once the electric field has been neutralized" (48). Point to note: Beckmann implies that gravitational and magnetic forces are identical; we'll see.

7. *On electromagnetic momentum*: Since the concept of an "elastic ether" has been abandoned and with it the possibility of a mechanical "stress" (like when a hammer hits a chisel) in a vacuum, the concept might better be replaced with "force density" (53). "Since force is the rate of change of momentum, it follows that a momentum . . . must be associated with an electromagnetic field. This phenomenon is . . . called, 'inertia of the electromagnetic field'" (54). A "magnetic field resists being changed." An "electric field resists being changed" (54). "Electromagnetic mass is no formal mathematical trick. It is a physical reality that a charged body resists acceleration *beyond* the resistance offered to it by its Newtonian mass" (55).

8. *The Lorentz contraction*: Beckmann asks, what happens to a "point charge" when the charge moves with uniform velocity with respect to a rest frame? (57). The principle of relativity dictates that the field, "must travel unaltered with the particle . . . otherwise we could . . . look at the distortion of the field surrounding the particle and without reference to any rest standard, we could proclaim with what absolute velocity the particle is moving" (57). "From this, (and certain facts concerning the Maxwell equations), Lorentz concluded that electrons contract in the direction in which they move," through what he thought was the "ether" (58). But "in reality there is no such contraction . . ." (59). In any event, "it was for this purpose that he introduced the transformation named after him, a very different purpose from that for which Einstein used it." This explains why the man who's discovery "provided the backbone of the Einstein theory remained irreconcilably opposed to it to his death in 1928" (58). Chalk up another great physicist who never accepted Einstein's theory.

    To Beckmann, the contraction of an electron as it moves through a dominant gravitational field "is no more a violation of (the principle of) relativity than rain drops being deformed as they move through the atmosphere" (Beckmann 1987, 59). But Einstein had to build the contraction into space itself and tie it to a subjective observer, "which in turn made the dilation of time inevitable" (59). The physical reason for this apparent contraction, having nothing to do with space and time distortion is explained by Beckmann below.

9. *On famous formulas and the relationship of mass to energy:* The formula for "the electromagnetic mass, associated with the resistance of the electromagnetic field to the acceleration of its source charge, had been known to the classics of the late 19th century . . ." The Einstein theory "derives the same type of formula for *any* mass, charged or neutral" (62). Another formula "imbued with a mystic romanticism it does not deserve . . ." shows that, "the quantity of matter does not . . . increase with velocity. What increases is the inertial reaction or resistance to a force changing a body's momentum" (63).

The most famous formula, $E = mc^2$, has fascinated laymen, often being "the only thing they know about the Einstein theory. But even some physics professors have romanticized 'the equivalence of mass and energy.' A glance at (its derivation) shows that this 'equivalence' is (an) absurdity" (Beckmann 1987, 64). When a body is discharged, the field, and its energy, "disappears only from its immediate surroundings: it is radiated away" (65). But the conservation of energy demands that the inertial mass be decreased by a corresponding amount, and Einstein is fully aware of this fact (65). This famous formula is also derived using classical physics only (63).

10. *IAAD, billiard balls, and electrons:* To show that Beckmann's proposed theory does not contradict the experimental evidence of Einsteinian dynamics, what needs to be shown is that *in all experiments whereby "v" was understood to mean velocity with respect to the observer, the observer was always at rest with respect to the local force field, meaning Earth's gravitational field* (73). One experiment, that of Champion, (1932) to test "mass-energy momentum, relations was widely interpreted as a confirmation of the Einstein theory" (75). But in Beckmann's opinion it doesn't measure up, though it does support his own theory (76). Billiard balls, which follow Newtonian conservation of momentum physics, "transfer momentum by actual contact at a point where simultaneity holds for all observers." But this is not true of an electron "because it does not wait in space, nailed to its coordinates . . . until it is bodily hit by another electron" (77). Instead, the electrons repel each other at a distance and continue to interact *at a distance* (77). But "the Einstein theory does not recognize the equality of action and reaction at a distance: the force exerted

by a particle at rest on a moving one (in the Einstein theory) is not the same as the force of a moving particle on one at rest" (77). Beckmann nevertheless has no doubt that through use of "opaque acrobatics of four-vectors and world lines" the Einstein theory can explain things as it usually does (77).

11. *Time dilation refuted*: There are three types of experimental "proofs" used to verify Einsteinian time dilation; the rate of decay of fast moving mesons in which the rate slows down; the slowing of accurate clocks transported around the globe; and the Doppler Effect. "What all three techniques have in common is the failure to ask . . ." or answer, the question, "is the measured effect something that is dependent on the observer, or something that changes the clock?" (77). Indicative of this was the famous Ives-Stilwell experiment (1938, 1941) to test the Doppler Effect of fast moving mesons (78). To Einsteinians it was considered proof of time dilation, but was it? This experiment is "impressive because its result depends only on a comparison of spectroscope readings, not on inferred velocities" (79). To Beckmann, "it is proof that particles traversing a gravitational field radiate, in their own rest frame, an inherent frequency. . . . This is a frequency an observer sitting on a moving particle would measure" (80). Ives apparently thought so, since, resenting the purpose to which his work was made to serve, he became a lifelong enemy of Einstein (21, 65). Chalk up another great physicist who never accepted Einstein's theory.

A more recent development which won't be publicized by Einsteinians any time soon is worth noting here. In the 1990's Tom Van Flandern, a physicist from the University of Maryland, (and one of our Principals to be addressed in another chapter), relayed the following story regarding the Global Positioning System (GPS), on which he worked as a scientific consultant. This is a series of atomic clocks placed in satellite orbit around the Earth. Bethel, who relayed the story, tells us that at an altitude of 20,000 kilometers, where the gravitational field is less dense, the clocks run roughly 46,000 nanoseconds a day *faster* than at ground level, but 7,000 nanoseconds *slower* than at rest, because of their orbiting speed of 3 KPS, for a net *increase* of 39,000 nanoseconds a day. The clocks have to be turned back by the engineers to compensate and thereafter keep perfect time. This is a refutation of Einstein theory which predicts that the clocks should run slower relative to observers, in practice, on Earth. But Einsteinians won't admit it (Bethel, April 1999). Van Flandern has also elaborated on this topic more than once in his Meta Research Bulletin, (e.g., Van Flandern, June 2003).

Beckmann reiterates his major premise in the rhetorical question: "Should physics seek to understand objective reality, or should it de-

scribe an observer's perceptions?" (Beckmann 1987, 81). This empha-
sizes again the struggle between Kantian subjectivism and objectivism in
science.

12. *Space-time for moving media*: Einstein's disciple, Minkowski, introduced the concept of "space-time" as opposed to space and time taken
separately, in his effort to apply the Einstein theory to moving matter. He
found the solution (1908), "via six vectors and their space-time components . . ." (85). But his equations "have never been verified to the second order" and his electromagnetics is "an esoteric, highly theoretical
field, which is not without problems . . ." (85). "On the other hand, for
slowly moving media . . . equations can be derived without the Einstein
theory, are verified, and do provide physical insight, for they rest on simple principles" (85).

13. *Things that stay the same at different observer speeds*: "The force between the plates of a capacitor, measured in uniform space and time,
cares very little about observers observing that force, even if they travel
past the capacitor at half the speed of light" (89). Even for "electromagnetic Doppler Effects and aberration, which *do* involve the observer's
motion . . . the effect producing velocity is that with respect to the local
(gravitational) force field . . ." (89). "There is no *a priori* reason why the
force mutually attracting two electric charges should have the slightest
dependence on the observer observing it" (89).

14. *Paul Gerber*: A famous "proof" of the Einstein theory, for which Einstein
is generally given credit for priority, was a formula predicting the ad-
vance of Mercury's perihelion. This indicates a slight change in the position of each revolution of all planetary orbits but is most pronounced in
Mercury's. It was derived by a high school teacher, Paul Gerber, using
only classical physics, when Einstein was nine years old (98). But Van
Flandern, although not an Einsteinian, agrees with Einstein's contention
that Gerber's derivation was wrong.

15. *Enough to whet one's appetite*: Sufficient evidence has been given to
support Beckmann's theory and to discredit Einstein, not as a charlatan,
for who can doubt his great genius and memorable personality? He will
perhaps remain an icon of the 20th century. Although physicists will hash
this out for some time to come since they can't be expected to change
their minds by mere philosophical arguments, and they seem in no great
hurry to accept Beckmann's first attempt to place the Einstein theory "on
classical ground." It should be clear, or at least open to reasonable doubt,
that Einstein's theory is based on Kant's subjectivist philosophy, and
leads into murky waters. If his theory was true, and there were no simpler or more logical alternatives, we would have to live with it. Ockham's

razor, which states that of two possible solutions, the simpler one tends to be the correct one, is after all just a rule of thumb. Still we should stay the course with reason, the law of identity, and objective reality, and see where it takes us. Beckmann has given us the lead, and he has much more. One hopes that science is still essentially objective and will self-correct when it has reason to do so.

16. *The Lorentz contraction explained and a glimpse at the nature of gravity*: In the second part of his book, Beckmann goes beyond Einstein and effectively contradicts him. Since two points dealt with there were mentioned earlier in this chapter, I will conclude by discussing them briefly. The first involves the "Lorentz contraction," important because it gave Einstein his "flash of insight."

Beckmann writes, "There is . . . a good physical reason for the charge distribution to change when it moves through an electric field, and that is a generalized skin effect" (Beckmann 1987, 150). Skin effect is a known phenomenon of sinusoidal or alternating current. But the same basic principle applies to the self-induced oscillations of an electron. This is explained in Part Two of Beckmann's book (see Chapter 8). "What matters is the skin effect is a genuine physical phenomenon that really happens, not a perception of an observer . . . as he travels past a charge" (151). It is "a genuine redistribution of charge as it traverses a field: a contraction (of the electron's field) in . . . Euclidean space that remains just as unaltered as when heat-shrink insulation contracts in it" (151, parenthesis added).

Second point: I wrote (see above, sec. 6) that Beckmann implies that gravitational and magnetic forces are identical. He now states, "Since all macroscopic matter (the only type where gravitation has been observed) is known to consist of positive nuclei and negatively charged electrons, the idea that gravitation is ultimately due to electric forces seems plausible . . ." (Beckmann 1987, 184). Even though there are flaws or obstacles to this theory, they are "no final refutation of the idea that the supposed twins (of gravity and electromagnetic force) may yet turn out to be one identical child" (186). Van Flandern, an astronomer who is also an expert on gravitational theory, has more to say on this subject (see Chapter 10).

Chapter Seven

Wigner's Friend: The Strange World of Quantum Physics

Science was born as a result and consequence of philosophy; it cannot survive without a philosophical—base. If philosophy perishes, science will be next to go.

–Ayn Rand

WEIRD SCIENCE

I believe the above sentiments to be valid—with the proviso that since the hard sciences are firmly rooted in the practical, and being more accustom to logical processes and more amenable to proof, they will not fall as easily as the social sciences. Nevertheless it is indisputable, as was discussed previously, that modern science was made possible by the liberating and rational philosophies of the Renaissance and the Enlightenment, and that science can not stand without them. It is also true that much of human history consisted of "dark ages" with respect to the advancement of science and knowledge in general. The present flowering of science and technology, it must be conceded, is rare in the annals of history. It is not an inevitable result of some automatic kind of "progress," which we are "guaranteed" and which can't be turned back. It is fundamentally linked to the *kind* of society we live in, and the *content* of its philosophical beliefs. Although present day Western Civilization is still predominately guided by philosophies favorable to science, this state of affairs is by no means permanent and unchangeable.

As concerns the present topic, the fact can not be ignored that exorcists, proponents of ESP and the "paranormal," communicators with the dead, and other mystics and charlatans, eagerly and opportunistically refer to quantum physics to validate their various arts. But could their claims be true nonetheless? It all depends on the verdict we pass on quantum physics—or more specifically, on the "weird" implications of quantum mechanics.

As is evident from the above and as we have been examining in previous chapters, I am attempting to show that science, and the things science gives us, is being threatened by irrational philosophies and beliefs—not to mention politics and the "easy money" of government funding. One is not being an alarmist to say so. There is ample evidence for this assertion. This is not to say that the trend is all bad, or that objective, practical, and empirical strains of endeavor are not still the dominant signature of modern science. Good work is obviously still being done. Indeed, the world has just recently witnessed the latest installment in scientific and technological innovation— namely the Information Revolution. But I want to stress again, that neither is it foolhardy to warn of the dangers and to point them out in some detail. This is what I propose to do as it pertains to the present subject.

IT USUALLY BEGINS WITH EINSTEIN

Albert Einstein, as we have seen in the previous chapter, was a man who, for all his genius, did more than any other scientist to tear physics from its firm foundations in objective reality with his "Theory of Relativity." He now plays the "man of reason" of the present drama. His statement, "God does not play dice with the universe," is well known. He meant that exact physical laws, and not indeterminate probabilities, should determine outcomes in physics. Whether he objected also on the grounds that quantum physics, or more specifically, *quantum mechanics*, (QM) violated the objective rules of science or the laws of logic is doubtful, since he himself did so, as I have tried to show.

This is nevertheless how the situation is portrayed in books on the subject. In famous debates with physicist Niels Bohr, the "weird" implications of QM continued to win out. The so-called "Copenhagen Interpretation" prevailed because the experimental evidence seemed to point in that direction. Even Einstein could not stem the tide.

THE COPENHAGEN INTERPRETATION

Any critique of quantum physics should begin with what it says. As is usual in the history of thought, it was a mixed bag; a mixture of the radical with the traditional, the logical and the illogical, the innovative and the weird. Physicist Alastair Rae, in his book, *Quantum Physics? Illusion or Reality?*, describes the turn of events within the context of *fin de siecle* classical physics. Rae does not qualify as a Principal in our story, but on the present subject he is for the most part, an honest broker.

At the close of the 19th century, writes Rae, few scientists believed that there were any new fundamental discoveries to be made. "But within thirty years a major revolution had occurred" in physics (Rae 2000, 2). Rae characterizes this revolution as follows:

1. Quantum physics, or more specifically, quantum mechanics (QM), leads to the rejection of scientific determinism. This is basically the doctrine of "cause and effect" in a reality governed by physical laws; as opposed to not only "chance," but also to outcomes that defy any reasonable understanding of the world as we know it. This idea will be discussed further below.
2. QM tells us that nothing can be measured without influencing and disturbing the measurement.
3. Thus the "Observer" becomes a crucial factor in the experimental process.
4. The paramount importance of the Observer is sometimes taken to the point of solipsism, that is, of believing that the Observer's mind is the only true reality and that the rest of the universe is an illusion or at best a probability. This is the meaning of "Schrödinger's Cat," as we will see below.
5. The implication is raised that ours is not the only universe, that other "parallel universes" are the best explanation for certain experimental results.
6. The postulate of other universes can restore the validity of determinism; effects have causes, though not necessarily in the same universe (3).
7. Quantum objects sometimes behave like particles, sometimes like waves. This is the so-called *particle-wave duality.*
8. The more accurately you know an entity's position, the less accurately you can know its velocity, and vise versa. This is the *Heisenberg Uncertainty Principle.*
9. The actions of an observer can affect what another observer sees—even at vast distance.
10. These "actions at a distance" occur instantaneously, or at many times the speed of light, though not in a way that can be used to send messages (Van Flandern 1993; Gribbin 1984).
11. A measurement of one property usually destroys all knowledge of another property of a quantum system. This is described by Bohr as *complementarity* (Rae 2000, 49).

12. We have to accept *non-locality* as an intrinsic fact of nature (53). This is the idea that an action here has instantaneous effects in remote other parts of this universe — or other universes.
13. As quantum phenomena assume ever larger proportions, approaching the macroscopic level, their laws begin to correspond to those of classical Newtonian mechanics. This is the principle of *correspondence.* At first glance this might seem like a common sense principle. But as we shall see in the next chapter, this too may be incorrect.

This revolution in physics is best known as the *Copenhagen Interpretation*, Copenhagen being the home of Bohr, its leading advocate. It was first published in a lecture given by Bohr in 1927. Although some of the more radical implications, such as multiple universes, had not yet been devised, and some of the best tools and experiments used as proof had not yet been proposed or made, the essential ingredients of Copenhagen were already in place by the 1930s and are still the dominant interpretation of QM today (Rae 2000, 48–53; Gribbin, 1984).

THE REAL QUANTUM PHYSICS

Proposing to critique QM as expressed in the Copenhagen Interpretation is not the equivalent of selling memberships to the "Flat Earth Society." Yet there are two reasons why such an epithet may be the likely reaction of many intellectuals and scientists. The first is that quantum physics may be, and often is, viewed much more broadly than the description given above would indicate.

In its broad interpretation, quantum physics is synonymous with contemporary or modern physics, as opposed to the "classical physics" of Newton and Maxwell. It is the era of physics which began with the unlocking of the secrets of the atom by men such as Rutherford and Thompson. It includes Bohr's model of the atom, the discovery of x-rays by Roentgen and von Laue, Planck's constant; and the discovery of the quanta or photon from the photoelectric effect by Einstein, and many other discoveries (Gamow 1961). There is no intent here to critique these discoveries nor would any such critique be justifiable.

The second reason to ridicule any calling of the aspects or implications of QM into question is the success record of the so-called "quantum cookbook" in bringing about many of the inventions of this present age of high technology. These inventions include, among other things, computer technology, specifically, components such as the semiconductor, the superconductor, and lasers (Gribbin 1984). But this is often a case of "apples and oranges." Statistical quantum formulations, even purely descriptive ones lacking in any real insight into the underlying causes, don't necessarily have to translate into "weird science." They are merely theories that work, without knowing the actual causes, as is often the case in science. Newtonian gravity is a classical example of a hard and fast law that doesn't concern itself with its cause. That issue will be addressed as we proceed, and in another chapter.

This chapter will attempt to avoid such broad strokes or package deals. I intend only to zero in on some aspects of the Copenhagen Interpretation, or its spin-offs. What I criticize, I will attempt to back up with evidence or some compelling reason. If a scientific theory is undeniable, but some or all of its implications are bizarre or assault the common sense, I will try to invoke Ockham's razor and find a simpler explanation, if one is to be found. If the theory itself is faulty or has some logical flaw, I will try to demonstrate that also. Yet this is no small thing. The Copenhagen Interpretation of QM, however strange seeming to the layman, has all or most of the weight of the scientific community on its side. Therein lies the danger which Ayn Rand warned against. Let the reader judge the success of this effort.

EPR AND BELL'S INEQUALITY

In his effort to prove that "God doesn't play dice," that the fundamental, deterministic laws of physics are not violated at the quantum level, Einstein, with co-workers, Podolski and Rosen, set up a thought experiment whereby information about one particle could be used to deduce properties of a second particle some distance away, such as its position or momentum (Gribbin 1984). This became known as the EPR or "two particle" experiment. But instead of proving Einstein's point, it was used as strong evidence that the Copenhagen Interpretation was the correct one.

Although the EPR was originally conceived of as a thought experiment, actual experiments have been done and perfected over the years. Instead of showing that we could deduce the properties of the second particle from the laws of physics, it was concluded that the very examination of the first particle *changed* the properties of the second particle instantaneously, or at many times the speed of light, though not by a method which could be used to send messages (Rae 2000, 45–6). Why can't this information be used to send messages? One answer is, *by definition*, because it would violate Einstein's "speed-limit," i.e., the speed of light. Another answer concerns the difficulty of understanding the possible "message." This will be explained as we proceed. In any case, the implication of this "non-locality," where information is transferred instantly to distant parts of the universe or to other universes, is bizarre indeed and goes to the heart of any philosophical objection to QM.

If faster-than-light communication is observed experimentally, but does not occur by conventional means, either by electromagnetic waves or by some other physical means, then that fact must necessitate a fundamental altering of our perception of reality. There are lots of questions and objections; but let's go into a little detail first.

As Rae explains it, "To understand the EPR problem we consider a physical system consisting of atoms—with the emission of two photons in rapid succession. The wavelengths of the two photons are different—but the most important property is that their polarizations are always at right angles, (90°)" (Rae 2000, 29, parenthesis added). We know that these polarizations of a certain type are always at right angles by quantum theory. But we also know it because this property can be directly measured using horizontal / vertical (HV) polarizers (Rae 2000, 29). Einstein's purpose essentially, was to show that some unknown or inherent property, what is now called the "hidden variable" theory, dictates what the second particle's properties will be in relation to the first particle (28).

In 1951, physicist David Bohm demonstrated that the point can be simplified by measuring a variable such as photon (light) polarization. Pairs of photons are emitted from a light source in opposite directions, and measured through HV polarizers. This is done by counting them and recording the "hit / miss" pattern with electronic detectors. If the left-hand polarizer is horizontal, the right will be vertical and *vise versa*. But if we omit the right-hand polarizer, how will we know what the result would have been if it *had been* there? Then, if we vary the angles of the photons passing through the HV polarizers, and take statistical measurements of large numbers of them, (since individual photons are not actually measured), will the results be predictable of the 90° variance, mentioned above, (which we'll call a "match,") between individual pairs of photons of each set? The hidden variable theory says "yes," but the dominant view, because the evidence seems to support it, is that of QM which says "no," that there will be other outcomes, those which conform to a "non-local" view of reality.

Although "indeterminism" and "uncertainty" are built into the very foundations of QM, (Rae 2000, 24), QM nevertheless has specific predictions as to the outcome of this EPR type experiment. Logical consistency aside, it is not unusual for a particular theory to be made to conform to a dominant philosophical view—such is the case with the Copenhagen Interpretation.

A basic hidden variable theory predicts that when the HV polarizers are in the horizontal and vertical positions, but offset by 90°, the statistical matches will show a perfect correlation, a correlation of 1.0, meaning all matches within some small margin of error. When the second, or right-hand, (RH) polarizer is rotated slightly by some angle, the correlation decreases slightly until at a 45°, there is zero correlation, meaning that any "matches" are purely random. Again, when the RH polarizer is placed in the same position as the LH polarizer, there is a perfect anti-correlation of -1.0, meaning all "non matches," again within some small margin of error. This is assuming that a photon always emerges through the polarizer channel whose direction is always nearest the actual polarization of the light (Rae 2000, 34). The QM theory, and actual experimental results, however, seem to show a very different outcome.

At this point the Bell Theorem or "Bell's Inequality" is introduced. It simply states that if three experiments are done of the type described above, each with the HV polarizers set at different angles, "the total number of pairs in which both photons are registered 'positive' in the second experiment, can never be greater than the sum of the numbers of 'doubly positive' pairs in the other two experiments," (Rae 2000, 40), provided that the results are determined by hidden variables, and are unaffected by "non-local" particles or measurements.

What's really being said here is that Bell's Inequality would be a logical necessity in the everyday objective world of classical physics. The fact that it is violated in QM proves that quantum theory has strange and wonderful rules of its own, unlike those of the world as we know it. I'm not being facetious here; this is the best way to explain what's really at stake. But the questions to ask are; 1) Does quantum theory really violate Bell's Inequality? And; 2) Is Bell's Inequality a valid tool for accessing quantum theory?

Let's look at this from another point of view. As Tom Van Flandern, who believes that Bell's Inequality is based on a logical flaw, explains it,

"We must regard the polarized light leaving the common source as pure waves, not as particles—at least until they actually hit the detectors. . . . For pure waves, there is a certain probability that the detection-miss pattern will be correlated at the two observers; and there is a complimentary probability that it will be *anti-correlated* at the two observers. It was the anti-correlation probability that was omitted in the reasoning which led to Bell's Inequality. When we consider the probability of anti-correlated events, it correctly predicts the experimental results—that the number of discrepancies will be more than double at 2 xs over what it was at small angle x.—And it does " (Van Flandern 1993, 126).

Thus the experimental results are explained in a logical way which preserve "locality." Whether the physicists conducting the various versions of this experiment could have committed such a glaring logical error remains to be seen. My hunch is that the power of a venerable philosophical theory such as the Copenhagen Interpretation, with the added impetus of peer pressure, along with financial and career implications, prompted them to just such a result.

WAVE-PARTICLE DUALITY

A "duality" implies a paradox or an apparent contradiction between two enti-
ties or attributes which defies normal modes of objective reality, and the Law
of Identity, (see Chapter 6). Ever since the famous "two-slit experiment" of
Thomas Young, (1800), light was known to be a *wave phenomenon* because
it was shown to unambiguously possess the qualities of *interference* and
refraction—qualities known to be characteristics of waves.

Thus Christian Huygens, an early advocate of the wave theory, won his
long debate with Newton, who favored the ballistic or particle theory of light
(Gamow 1961). With the advent of QM, however, the controversy again arose
but this time in a different form. Now light, or any wave on the electromagnetic spectrum, was believed to be *both a wave and a particle*, if not at the
same time and in the same place, then depending on how it was measured.

Again it begins with Einstein. In this case, it was the photoelectric effect
which led him to his theory of light quanta or photons, as noted above. When
light is directed onto a piece of metal or other photosensitive material, an
electron is knocked out and can be detected by applying a voltage in the correct manner. The energy of the emitted electron does not depend on the
brightness of the light but on its frequency or wavelength (Rae 2000, 6;
Gamow 1961). Further, photons seem to bounce off electrons, conserve energy and momentum, and behave like particles rather than waves (Rae, 6).

But if photons are passed through two slits using very weak light, and even if
they pass through the slits almost one photon at a time, they nevertheless exhibit
interference patterns just as if they were waves (Rae 2000, 7). But if one slit is
blocked, or a shutter is moved back and forth over the slits, so that photons can
positively only pass through one slit at a time, the interference pattern is gone!
Thus, "the fact that processes like two-slit interference require light to exhibit
both particle and wave properties is known as *wave-particle duality*" (9). It
should be noted here, in fairness to Rae, that although he ultimately favors the
Copenhagen Interpretation, he also presents opposing views clearly enough to
allow the reader to judge the facts on their merits to some degree. This cannot be
said of Gribbin and other writers on this subject.

It should also be made clear at this point that although there are theories
which offer more rational explanations to so-called wave-particle duality,
such as the theory which instead of presenting waves and particles as alternatives, posits them as occurring simultaneously, with the wave having the
function of guiding the photon along its path. But these and other hidden-
variable theories are not supported by the experimental evidence in all cases,
(Rae 2000, 26–7), as is apparently, the Copenhagen Interpretation.

So what's wrong with wave-particle duality? Petr Beckmann, who will
presently have something to say about the subject, maintains that QM is more

given to *description* than to *insight*. According to the prevailing interpretation, a photon passing through a two-slit apparatus behaves like a wave because it *is* a wave—even though there is good evidence that photons can not
be split. When it is detected by a photographic plate it behaves like a particle
because it *is* a particle (Rae 2000, 50).

Is this the nature of "deep reality," even though QM holds that there is no deep
reality? Or is this merely a convenient explanation of observed phenomena,
made possible by Kant's, or more specifically Berkeley's philosophy?

Beckmann sees it differently. After demonstrating how key elements of
"contemporary physics" can be derived form the Maxwell equations, and
from classical physics, "simply and without further assumptions," Beckmann
explains,

> "The fact that it is the Faraday field that obeys the Schrödinger equation allows
> a resolution of the two-slit paradox and refutes the 'wave-particle dualism. . . .
> A wave is incompatible with a particle for at least three reasons . . . 1) a wave
> can be split into two or more parts; 2) it does not repel or attract other waves;
> and 3) it is attenuated, even in a lossless medium, by natural dispersion, such as
> that governed by the inverse square law.
>
> "If ψ is a standing wave with respect to the moving electron, then clearly this
> wave will pass through both slits, though the electron passes through only one. It
> can be shown . . . that the electromagnetic waves re-radiated by the two slits will
> slow to the group velocity in the neighborhood of the electron, which itself be-
> comes the prisoner of that field. The position of the electron striking the screen
> is then determined by the interference field of the two electromagnetic waves,
> which have been slowed to have a de Broglie wavelength rather than a free space
> wavelength—just as observed in experiments "(Beckmann 1987, 154).

What Beckmann is saying here is that the so-called "wave-particle duality," is
no duality at all. It is another case of "apples and oranges," or more specifically,
of *electrons and electromagnetic waves*. Electrons have certain properties which
can be measured, and so do waves. Electrons have standing, or carrier waves,
and when excited they radiate electromagnetic waves. But *the question is
whether these waves themselves are also particles*. Beckmann says they are not.
His book was, for the most part, ignored by mainstream physics.

A WEDGE OF UNCERTAINTY

Most of the developments in QM, beginning with Plank's constant, (in the
last week of 1899), ranging all the way to de Broglie's theory of "carrier
waves" for the electron (1924), and Schrödinger's equations (1926), which essentially quantified de Broglie's theory, were in the spirit of a logical, if not

deterministic physics, albeit that they relied on statistical formulations. That changed with the introduction of the Heisenberg Uncertainty Principle (1927).

The Uncertainty Principle too begins in a logical, if not a deterministic way. It begins with the observation that any attempt to apply the ordinary methods of observation and measurement to quantum phenomena is bound to end in failure (Gamow 1961). Take the simple example of measuring temperature. If you stick a thermometer outside your window, you expect the thermometer to measure the temperature of the air surrounding it. But suppose the thermometer had previously been in your refrigerator. You don't expect the cold thermometer to *lower* the surrounding air temperature. In fact it does just that, though in such an infinitesimally small amount that it is regarded as insignificant.

Not so in QM. Although scientific experiments always have to take into consideration the fact that the instruments used in an experiment may somehow influence the outcome in ways we do not want; in quantum physics, the problem becomes much more pronounced and exaggerated because we are dealing with the smallest known objects. Thus, for example, the very act of *observing* an electron changes the electron's behavior, because the light waves or photons used in the observation cause the electron to do something entirely different from what it might have done in the absence of the observation—hence the uncertainty. We can not be certain, said Heisenberg, of observations or measurements at the quantum level.

This obvious problem was the wedge which pried open the lid of Pandora's Box to the bizarre world of QM as manifested in the Copenhagen Interpretation. There was no particular villain; irrationality was—and still is—in the air, made possible by philosophers like Berkeley, Hume, and Kant, who needed only an opening to infest the world of science. Einstein gave Kant his opening, as discussed in the previous chapter. Bohr gave Berkeley his.

How was this leap of logic from the rational to the irrational made? Gamow, who was a student of Bohr, and was thus around when all this was happening, spells it out for us.

"The energies on this scale are so small that even the most gently performed measurement may result in substantial disturbances of the phenomenon under observation. . . . The observer and his instruments become an integral part of the phenomenon under investigation. *Even in principle there is no such thing as a physical phenomenon per se*" (Gamow 1961, 255, Italics added).

I do not wish to belabor the philosophical issue. I only want to point out that such a logical leap can not materialize out of thin air. There is no other explanation for going from the common sense idea that the smaller things become, the harder they are to observe and measure, to the *fantastic* claim that "*in principle there is no such thing as physical phenomenon per se!*" If that were really true, we might as well not have science at all. But that is just what happened in the world of quantum physics, and is the dirty little secret behind the Copenhagen Interpretation.

It is not my purpose here to try to disprove Heisenberg's scientific argument. But Carver Mead, and again, Beckmann, will have something to say about that in the next chapter. Heisenberg said in effect, that we can't track an electron's position and momentum at the same time because of the problem mentioned above. He did, however, show a way to average the two measurements in order to achieve one fairly accurate measurement—precise enough in fact, for cathode-ray tubes and television to be developed, where the "exact" tracking of an electron's position and momentum becomes extremely important.

The interesting point for our present purposes is to show that this led to the increasingly widely held belief that physics is ultimately governed by *stochastic* (meaning governed by probability), rather than *deterministic* laws (Gamow 1961). But this is not a theory of "chance" or probability as it is usually understood. In conventional probability theory, for example, if we knew enough about the random or chance flow of air molecules and weather patterns, we might be able to predict the weather in an exact way, long into the future. We might even be able to predict global warming! In the new theory, we can never know the exact position of electrons, or any elementary particles at the quantum level, or even if they *have* an exact position—or even if they *existed* before we measured them.

At this point Niels Bohr embraced Heisenberg's principle and turned it into a philosophy of physics—the Copenhagen Interpretation—as discussed above. And Einstein made an unsuccessful attempt to combat it, also as mentioned above.

In the next step, Dirac united quantum theory and Einstein's relativity theory, and predicted two different worlds, one positive and one negative! (Gamow 1961). From there, it was a simple step for physicists to predict multiple universes. I'll interject here, for the sake of clarity—and sanity—that although subsequent experiments confirmed the existence of positively charged electrons, or *positrons*, even the probable existence of *antimatter*, this in no way proves "two worlds," or multiple universes. It may simply be that different particles at the quantum level have different properties just as they do on our scale.

WIGNER'S FRIEND

The symbol of what quantum physics has become is exemplified by "Schrödinger's Cat." This is the popular name given to Schrödinger's famous thought experiment. As Gribbin explains it in his book by that name:

> "Imagine a box that contains a radioactive source, a detector, (a Geiger counter), a glass bottle containing . . . poison . . . , and a live cat. . . . There is a fifty-fifty

chance that one of the atoms in the radioactive material will decay and that the
detector will record a particle. . . . Then the glass container is crushed, (by a
hammer), and the cat dies" (Gribbin 1984, 203, parenthesis added).

Since radioactive decay is believed to be a purely random occurrence, there
is no way to know exactly *when* such an occurrence will take place—except
statistically. But, according to the Copenhagen Interpretation, not only do we
have no way of knowing whether the event has occurred, but we can say, in
all seriousness *that it has not occurred at all until it has been observed.* Hence
the central position, and singular importance, of the "Observer" in QM. The
bottom line—and the paradox—is the belief that *the cat is neither dead nor
alive until she is observed to be one or the other!* (Gribbin 1984, 205).

Gribbin unabashedly defends this along with all the other weird manifesta-
tions of QM in his book, and again in a sequel called *Schrödinger's Kittens*
(Gribbin 1995). He does so in the name of hard science, backed up by the experimental evidence. And he has no apparent problem with the logical contradictions and absurdities. On the other hand, as was mentioned above, Rae
does have problems with the weird implications of QM, although he ultimately leans toward Copenhagen, he at least gives reason a fair hearing.

Through a contorted process of reasoning, the distinction between what can
be accepted as factual, reality based evidence in QM, and meaningless conjecture, is the presence of a conscious Observer, by this it is usually taken to
mean a human Observer. Hence the cat may not know if she is dead or alive—
and more importantly—would not *be* dead or alive until an Observer looks to
see, but surely a human Observer inside the box would know—and would be.
This, at least, was the idea of E.P. Wigner and his "consciousness-based theory of measurement." Wigner's theory draws a fundamental distinction between human consciousness and every other entity in the universe. Thus when
his human "friend" in the box makes her report, you can believe it! More importantly, you can believe that she really existed while she was in the box!
(Rae 2000, 64). I for one also believe that there is something special about human consciousness, but that doesn't make it a scientific fact, and especially not
the fact which determines the existence of the rest of the universe.

Rae relies on the philosophies of Carl Popper and Bertrand Russell to support his skepticism, but without an objective world view, supported by hard
scientific evidence and scientists who refuse to accept the logically impossible, he is bound to remain indecisive.

Other variations of the HV polarizer experiments have lead theorists to
conclude the existence of "many worlds" or "parallel dimensions," perhaps
thousands, or millions, all different from ours in some small respect. Add to
that the possibility of travel in time—again supported by the evidence, and

you have the state of the art in quantum physics today. Anything to keep from seeing the hum-drum of objective reality—which is really not so hum-drum at all, when you look at it as it really is.

So now, the reader can watch re-runs of "Quantum Leap," and "Sliders," with the full assurance that they are not far-out sci-fi, but instead represent the cutting edge of theoretical quantum physics.

Chapter Eight

The Electricians: Collective vs. Galilean Electrodynamics

INSIGHT WHERE CONFUSION HAS GONE BEFORE

To the mind that delights in magic formulas and incomprehensible mysteries, quantum physics and quantum mechanics as described in the previous pages holds a strange fascination. But the logical mind searches for something more. How is the world really made? Is it bizarre beyond all recognition at bottom, or is it more like the world we see around us—the world of common sense? We may never know, but there are those of us who are not ready to give up the quest just yet.

This chapter, and others to follow, will deal with scientists who are trying to answer these questions. But they are not the scientists who give us the familiar, commonplace theories, or even those bizarre new theories, such as String Theory, which are acceptable to the mainstream, and taught in the colleges and universities. These are the Apocryphal thinkers who offer new theories in keeping with the theme of this book—theories which can be validated by the evidence, and theories which are logical, but for one reason or another are at odds with the established paradigm. Some of these thinkers may be the Galileos and Faraday's of the future.

This chapter examines two scientists' positions on quantum physics and quantum mechanics, which to a large extent demonstrate that they can be understood by employing concepts and modes of thinking familiar to electricians—and to that extent they are familiar notions. As before, I am concerned with the conceptual content of the theories presented as made evident by the qualitative descriptions of the phenomena given by their authors. One assumes with good reason (mainly because of their authors' professional credentials), that they both got the math right, and if there are doubts, that is less important than their insights.

Mathematics, although necessary to the formal theory, can be used to prove almost anything. Since one of the goals of this book is to help the general reader to actually understand what the scientists are saying, we will proceed as before, in words any intelligent layman can understand.

Two physicists, both professors of electrodynamics in their time, will present their ideas here. Both have tried, in their own way, to give insight where confusion has gone before. They are Carver Mead, and once again, Petr Beckmann.

CARVER MEAD TAKES A STAND

Mead's idea in a nutshell is that certain electromagnetic phenomena, in particular superconductors, represent "collective" quantum systems or "huge atoms," if you prefer. Thus quantum phenomena, which once required elaborate statistical models, such as the Bell Inequality / EPR type experiment discussed in the previous chapter, can now be studied collectively. In this way, precise deterministic laws, as opposed to the statistical laws of QM, can be formulated. Mead, a long time professor at Caltech, is a well known and highly respected pioneer in the computer revolution of the past thirty years. So his ideas will not be able to be easily dismissed, even by the most entrenched and stagnant part of the mainstream; but that they will try, I'm sure.

In Mead's book, *Collective Electrodynamics, Quantum Foundations of Electromagnetism*, he states that the past seventy years will be considered as "dark ages" in the history of theoretical physics. It was for this reason more than any other that he was included among my Principals. He writes:

> Statistical quantum mechanics has never helped us understand how nature works; in fact, it actively impedes our understanding by hiding the coherent wave aspects of physical processes. It has forced us to wander seventy years in the bewilderness of "principles"—complementarity, correspondence, and uncertainty. . . . Correspondence to classical mechanics was the worst cause of conceptual nightmares (Mead 2000, 123).

But Mead's heroes, and his models of what a physicist should be, are not such classical physicists of electrodynamics as Faraday and Maxwell. They are Einstein and Richard Feynman—the king of modern physics, and the lovable and famous professor of physics at Caltech who preceded Mead. Both were in the epicenter of the contemporary physics mainstream, but both anticipated his idea that quantum mechanics could be understood conceptually, in a precise, deterministic way. To that extent, both opposed the Copenhagen Interpretation. It all goes back, once again, to Einstein's battle with Bohr over

whether or not "God play's dice," discussed previously. Mead sees Einstein
as the ultimate winner of that debate, even though science had not progressed
enough at the time for Einstein to realize it.

CORRESPONDENCE—TO WHAT?

According to Bohr's *correspondence principle,* quantum laws are uniquely
different from mechanical laws until the quantum systems become very large,
or macroscopic, whereupon they then begin to resemble, or *correspond* to
Newtonian mechanical laws of physics. Mead sees this principle as entirely
mistaken. Large numbers of quantum elements begin to interact in a collective, *coherent* manner consistent with the laws of *electrodynamics*, not mechanics. By understanding electrodynamic phenomena in its purest forms, in
such devices as superconductors and lasers, we can understand quantum phenomena in an exact and certain way. Mead thus challenges in one stroke, two
pillars of QM; *correspondence*, and *uncertainty*.

The reader will remember from the last chapter that through experiments with
HV polarizers, and through "thought experiments" such as "Schrödinger's Cat,"
the conclusion of QM was that there is no such thing as a complete or exact description of individual particles, phenomena, or systems—*or even whether they
exist at all*—at least not in this universe. We can only understand them in a statistical way, (see Chapter 7). But Mead, following Einstein, rejects these "un-
natural theoretical interpretations" (Mead 2000, 3). "Classical mechanics is an
inappropriate starting point for physics because it is not fundamental. . . ." it represents rather "an *incoherent* aggregation of an enormous number of quantum
elements" (4). The only *coherent* systems in the macroscopic world are electromagnetic, because the wave functions of the elemental particles are in *phase*,
whereas in mechanical systems they are *incoherent* because the wave functions
are *random*. Referring to collective electrodynamics, Mead writes, "Nowhere in
natural phenomena do the basic laws of physics manifest themselves with more
crystalline clarity" (7).

SUPERCONDUCTORS AND MOMENTUM

While a detailed review of Mead's proofs is beyond the scope of this book, a
look at some of his insights and principles may help us to judge the merits of
his theory, from our point of view. A few facts will be noted as we proceed.

Mead states (Mead 2000, 10) that if we connect the ends of a supercon-
ducting loop of wire, and reduce the voltage to zero, the current will flow

indefinitely. This is called a *persistent current.* But if we open the loop and connect it to an external circuit to do work, "we can recover *all of the energy.*" Further, (13), he states that quantization can be understood "as an expression of the single-valued nature of the phase of the wave function." With the two ends of the loop connected to each other, the two phases must match up. Thus, a wave that is confined to a closed loop has a single valued and continuous phase. The collective electron system represented by the wave function is made up of such elemental charges in continuous phase.

Feynman had said that force as a concept is irrelevant in a quantum context. But we can make a connection between the force which drives electric motors and the underlying quantum reality which causes it through the concept of *momentum* (Mead 2000, 20). In a collective electron system, "the momentum is proportional to the velocity, as it should be. It is also proportional to the size of the loop, as reflected by the inductance. . . ." So the inertia of *each* charge increases linearly with the total number of charges, but the inertia of the *coupled* or collective charges increases geometrically, or as the square of the number of charges (21). After expanding further on details of forces on currents, Mead concludes: "We can see how the force law discovered in 1823 by Ampere arises naturally from the collective quantum behavior" (23). And concerning the nature of inertia in electric coils, Mead writes:

> The total inertia of the electron system in the magnet is much larger than the actual mass of the atoms making up the magnet. It is curious that the electromagnetic momentum has been largely ignored in introductory treatments of the subject . . . the momentum of the collective, interacting system is overwhelmingly larger than that calculated by adding the momenta of the free particles moving at the same velocity . . . people often speak of the "momentum of the field" instead of recognizing the collective nature of the system (25).

We will see shortly how these insights into the quantum nature of matter as seen from the vantage point of electrodynamics, were anticipated by Petr Beckmann twelve years earlier. I am not implying any question of priority here, merely giving credit to Mead for coming to similar brilliant insights and taking them further.

RELATIVITY AND LIGHT CONES

Mead now incorporates into his Collective Electrodynamics, Einstein's Theory of Relativity, (Special and General Relativity as needed), through the concept of four dimensional space, or "four-space" as it is called, and other mechanisms. He leads us through the canon of the history of physics, to show

how this interpretation is inevitable, indisputable, and even non-controversial (Mead 2000, 32–35). Through four-vector mathematics, he derives approximations to further develop his laws of Collective Electrodynamics.

As do many tracts on this subject, he buttresses this discussion of Relativity on two pillars: The Michelson-Morley experiment, which "proved that there was no ether," and the Lorentz transformation, which was used to demonstrate that space "contracts," (see Chapter 6), and was later used to demonstrate the mathematical possibility that light (or electromagnetic propagation) could be "symmetrical" in time. This led to the development by Einstein, Feynman and others, of the "light cone," a hypothetical construct which showed light emanating out into the universe in two directions—both forward and backward in time! This "symmetricality of time" is then also part of Mead's attempt to make quantum physics logical and comprehensible. Writes Mead;

> "The English language statement is that the four-potential at any point in space-time is the sum of all four currents on its light cone . . . all directions along the light cone are treated equally. In particular, there is no distinction between the "retarded potentials," generated by currents in the past, and "advanced potentials," generated by currents in the future (Mead 2000, 75).

Interestingly, neither Michelson nor Lorentz, who were both great physicists in their own rights, ever accepted Einstein's Theory of Relativity (see chapter 6). The fact that there is a growing body of knowledge, backed by evidence, which refutes this and other sacred canons of modern physics, may possibly be unknown to Mead. But if so, it's unfortunate. In any event, Mead's contribution to Apocryphal Science is substantial, and we will have more to say about it below.

In fairness to Dr. Mead, I now relate an excerpt from his response to the above criticism:

> The entire reason I included a discussion and clean derivation of the (Lorentz transformation) is precisely to show that it is an observer issue. So in this case Petr and I are in complete agreement. In fact, I dislike the warped space-time view of physics, and do not subscribe to general relativity for the same reasons I don't like the Maxwell equations (from a correspondence, *parenthesis added*).

THERE IS ONLY ONE PHYSICS

In Part Two of Beckmann's book, *Einstein Plus Two*, (which was not addressed in chapter 6), Beckmann deals with issues which have a direct bearing on quantum physics (and QM), and which anticipate Mead's concept of

Collective Electrodynamics. His central purpose is to show, that all of the pillars of QM, including *Plank's constant*, the *De Broglie hypothesis*, and the *Schrödinger equation*, can be derived without the use of Special Relativity (SR). But he even goes Einstein one better to show that they cannot be derived by SR at all, but they can be derived, with insight, through classical, or what he calls, "Galilean" electrodynamics. A brief review of Part Two, (entitled "Einstein Plus One"), follows:

"In the case of the *collective* flow of electrons," said Beckmann twelve years before Mead's book came out, "no one doubts the phenomenon of self-inductance: as current increases, a magnetic field builds up, (which) induces an electric field which opposes the increase in current" (Beckmann 1987, 108, parentheses and italics added). It follows that this build up, by Faraday's law, induces an electric field which has a direction opposite to, and caused by, the acceleration (Lentz's law), thus reducing the magnetic field and inducing the electric field, and then accelerating the electron again.

The result is an oscillation with a certain frequency (Beckmann 1987, 108). "The natural frequency of these oscillations . . . must be an integral multiple of the orbital frequency, or the orbit will not be stable." The result is a discrete set of orbits around an atom's nucleus; *Bohr's orbits* (109). If an electron, or a body, is charged, when it accelerates it has to overcome not only *mechanical* inertia, (Newton's second law), but also *electrical* inertia, (Faraday's law). The frequency of oscillations is derived from a simple rule which Beckmann calls the "carrot formula," from which he derives, through straightforward classical electrodynamics, the *De Broglie relation*, *Schrödinger's relation*, and quantum mechanics, without the use of black-body radiation or atomic spectra, which were the basis for the statistical methods by which QM was derived historically, without ever knowing the underling causes (111). The reader will remember that Bohr said there *were* no underlying causes, and the world believed him.

The accelerated charge and the principle of relatively combine (this summary is of course simplified), no longer guaranteeing that the charge will move frozen to the field. The charge will catch up with, or lag behind, the equipotential, causing a higher gradient or contraction on the forward side, and a lower gradient to the rear. This is the famous Lorentz contraction (Beckmann 1987, 114, 137), and as the reader can see, it has nothing to do with a "contraction of space-time." In reality, it is akin to the skin effect of alternating current (150). Thus the electron oscillates or "wiggles" in its orbit, but the electron must have an integral number of "wiggles," or wavelengths per orbit, which is equal to the orbit's frequency or some multiple of it (122). The reader may notice the familiar ring this statement has to Mead's Collective Electrodynamics described above.

If we calculate the force and the mass of an accelerated charge, Beckmann says we'll find out something interesting. As already noted, the force is asymmetrical, the force on the leading side is larger than on the trailing one (Beckmann 1987, 138). The field mass is a full fledged mass that appears in the expression for force, momentum and energy. "In the case of energy, it is called *magnetic* when associated with a . . . charged body, but *kinetic* when associated with Newtonian, uncharged matter" (139). When calculated, it is shown that the total mass of an electron is *twice the size* of its Newtonian mass, because that mass and the Faraday mass are equal (135). The same exact rules would apply for a charged tennis ball, though the proportions would be vastly different. There is only one physics (136).

The principle of the way an electron orbits in an integral number of wavelengths or wiggles, can be demonstrated by turning a transmission line back on itself, or making a wire into a loop, and cooling it below the superconductivity level to make it lossless. By using a four-pole with lumped circuit elements so that it will have a matched impedance, and energizing it, it should have an electromagnetic wave going round and round—"lasting hours, perhaps even days or weeks" (Beckmann 1987, 157). Or as Mead explained it, it will last indefinitely.

A QUANTUM CHALLENGE

Mead seems to be saying that forward and backward time symmetricality is the most economical explanations for the EPR, Bell Inequality type experiments discussed in the previous chapter. He, with Einstein, maintains "that time can run in both directions in physical law . . ." (Mead 2000, 79). But one directional time is the necessary back-bone of the principle of *causality*. In order for a thing to have a cause, *some event had to precede it in time.*

Many physicists, including Mead, believe that causality has its roots, and is best demonstrated, not by fundamental laws of physics, but by *thermodynamics* (Mead 2000, 79). According to the second law of thermodynamics, which involves *entropy*, there is a gradual, one-way, irreversible loss of energy in a system—or the universe. This, according to Alastair Rae, (Rae 2000), poses a serious obstacle to the possibility of two-directional time. But the jury is still out. This area is still on the frontiers of physics, and the current reasoning may be questionable on many counts, as we will see in a future chapter.

So I leave this look at Mead's attempt to put quantum physics on a more rational footing. Did he succeed? In my opinion the answer is yes, in part, even though, as I have shown, he lends credence to some of the most vexing and difficult aspects of modern physics to reconcile with a common sense view of the world. Could they be true nevertheless? Of course, but we will proceed according to our plan and let the reader decide.

Chapter Nine

See No Evil: Heretic Arp's Challenge

THE NEW SCHOOLMEN

Not since the time of the Medieval Scholastic philosophers—the "School-men," as they used to be called—has an established hierarchy of scholars felt so absolutely self assured in the rightness of their beliefs and their power to impose them on the world; and never since that time has such smug complacency been so utterly and fundamentally misplaced, as is the case with modern observational astronomy.

Although much of the old Scholastic body of knowledge was really thinly veiled theology, it is easy to overlook the fact that much of their work was also considered the "hard science" of the time. For over a thousand years, well into the nineteenth century in fact, countless millions of people died needlessly as a result of their "medical science." In astronomy, which could hardly be distinguished from astrology, the intricate and finely detailed theory of "epicycles" was based on the theory that the Earth was the center of the universe. Although it clearly and "scientifically" explained the movements of the planets through space in a beautiful equivalency, it was thoroughly unfounded in fact. In both these cases as in many others, the arcane and fiercely defended "science" of the Middle Ages with its sophisticated "doctrines," proved to be utterly and completely wrong.

Today only philosophy students and historians read the old Scholastics. With few exceptions their "knowledge" has fallen into oblivion. But those who, after a long and hard fought battle, finally broke their death grip, are revered as the fathers of modern science.

Although there will always be those who think they have a monopoly on "the truth," today the type of philosophical / theological "truth" once represented by the Schoolmen is no longer considered to be science. And yet no modern *scientific* discipline more closely resembles Scholasticism as does modern mainstream astronomy, in its stubborn insistence on clinging to a set of supposed "truths" which have been thoroughly refuted by the observational evidence.

Today's received wisdom has come down to us from some of the most venerated names in modern science. Once again there is Einstein, who, through his theory of General Relativity, postulated that the universe must expand in order to prevent it from collapsing back into itself. And there is the astronomer Edwin Hubble, who discovered that the most distant objects in the universe produced a *redshift*. These objects were mainly the other galaxies outside our own Milky Way galaxy, then visible to the less powerful telescopes of his time. Hubble discovered that these distant objects were more redshifted then were local stars in our own galaxy, and postulated that they must be moving away from us at increasing velocity with distance. For lay readers not familiar with what this means, a *redshift* is a phenomena which results from the Doppler Effect (mentioned in ch. 6). Just as a train produces a varying pitched sound as it passes you and moves away at rapid speed, so too does light. As the object producing it moves away from us at rapid speed, the light waves reaching us shift toward the "red" end of the visual spectrum (i.e., the waves get longer); thus the term *redshift*.

From these beginnings, it was soon theorized by Friedman, (Arp 1996, 225), and others, that since the universe is expanding, it must have begun from nothing at a single point in time and space—before there was any space. A terrific explosion or "*Big Bang*"of this "nothing" produced the universe as we know it, around fifteen billion years ago, and it has been expanding ever since. From this it was also theorized that their must be "*black holes*," huge collapsed stars perhaps, with masses so large that even light cannot escape from them. Were they doors to another dimension perhaps? These must be "*singularities*," centers of "nothingness," like the one the Big Bang came out of. There must also be "*dark matter*," matter which can neither be seen nor detected, but which must be there, in order for Einstein's gravity calculations to work.

Then, *quasars* were discovered, immensely bright objects with extremely high redshifts. This meant that they must be both the most distant objects in the universe, and by far the brightest. What were these mysterious objects? As these and many other strange and mystifying phenomena were discovered, of a type which the old Scholastics might have approved, a new dogma arose — a new canon. This became the new received wisdom of establishment science. This is the "hard science" of today's Schoolmen.

The Principal of the present chapter, Halton Arp, says this theory—this paradigm—is wrong, and he has the evidence to prove it. Not surprisingly, he is being treated like the old Schoolmen treated Copernicus. This book would be useless if it were merely about what I believed and felt; just another addition to Scholastic literature perhaps. But it is more than that; it is more about the evidence of credible Principal scientists, coupled with their logic, than about my opinion. Here is some of Halton Arp's evidence.

Although Arp's ground breaking book, *Quasars, Redshifts and Controversies*, first introduced his new theory, we will review his more recent and more comprehensive book, *Seeing Red: Redshifts, Cosmology, and Academic Science*.

A GROWING BODY OF EVIDENCE

"In 1948, John Bolton discovered radio sources in the sky," writes Arp. They were coming from other galaxies outside our own. They tended to be in pairs, and radio filaments (signals which trailed from one radio source to the other), were found to be connecting the pairs. "It became clear that (the other galaxies') centers were the sites of enormous, variable outpourings of energy" (Arp 1996, 4, parenthesis added). But astronomers refused to believe that the radio pairs, or the X-ray pairs associated with them, were ejected from the galaxies they are connected to. To this day still refuse to believe such mounting evidence if it is accompanied by high redshifts (4). The reason for their refusal, like the Schoolmen of old, to believe their own eyes, "is the now-sacred assumption that all extra galactic redshifts are caused by velocity and indicate distance." You see, many of the radio sources had a much higher redshift than the galaxies they were obviously connected to; and that meant that, according to the mainstream, Big Bang theory, they *had to be coming* from much farther away; therefore they *couldn't* be connected (5).

Years later, in 1963, when these radio sources were being studied spectroscopically, (this is the way scientists analyze the electromagnetic waves emitted from objects), it was discovered that their redshifts indicated that the objects were receding at close to the speed of light. This *had to* indicate great distance, and that meant these objects *had to be* 10,000 times brighter than any previously known objects. As noted above, these were called *quasars* (Arp 1996, 5). Mainstream astronomers made this fantastic assumption, rather than believe their own eyes that the quasars were coming out of relatively nearby galaxies, for that meant they would have to challenge canon law—the Bing Bang theory.

Halton Arp suspected the truth early, and saying so, began to be considered a nuisance, or worse, a "crackpot," which is a modern word for "heretic," apparently. And now, 30 years later, when the observational evidence is becoming very much more clear, "there is a relentless effort" to ban the "new observations from conferences and suppress them from publication" (Arp 1996, 6).

So Arp went it alone, and published his findings. Some say he was wrong to buck the establishment, that he hurt his own cause, that another more subtle way could have been found. The best way to challenge any kind of ignorance or evil, intentional or otherwise, is always a difficult question. The important thing is that one does something.

A CASE IN POINT

The case of Markarion 205, illustrates the problem beautifully, a problem which repeats itself countless times throughout Arp's book. Each is considered an "isolated case" by the establishment, needing more proof, more supportive evidence. Meanwhile, as the evidence piles up and forms a mountain, the pattern becomes clear.

As a member of the Max Plank Institute (MPE), in Germany, Arp submitted a proposal to do a small project using the new orbiting, German engineered, X-ray telescope, ROSAT. Since the telescope had been launched by NASA, the US had 50% of the observation time, and there was little left for Arp's projects. However, since he was a veteran with long experience (Hubble had given him his first job), and still had a decent reputation in some circles, he was given some time.

> The proposal was to see if the connecting bridge from NGC4319 (a violently disrupted spiral galaxy), to Mark 205 (a quasar-like object), showed up in X-rays. As is often the case, the major aim failed. . . . *But what did show up was two X-ray filaments coming out of either side of Mark 205 and ending on point-like X-ray sources.* . . . I immediately got out the (existing) sky survey prints . . . to see if they were optically identifiable. Lo and behold! (They were). Of course they were quasars (Arp 1996, 19, parentheses added, italics in the original).

It turns out that these high redshift quasars had been observed before, but had been dismissed as insignificant through some bizarre logic (Arp 1996, 20). But the "stunning aspect of the ROSAT observations was that the two quasars of (high) redshift are actually linked by a physical connection to a low redshift object. . . ." (20, parenthesis added). I emphasize again that this flies directly in the face of the orthodox, accepted theory, and cannot be accepted by mainstream science unless it first admits it had been wrong all along.

What was the reaction of the "referees" when Arp submitted his findings for recognition in mainstream publications? (He doesn't say which one in this case, but mentions *Astronomy and Astrophysics*, in a similar one). The "experts" attributed his findings to "noise," and to "something wrong with the instruments," though Arp easily demonstrates that this was not so (Arp 1996, 20).

Another isolated case of noise, and faulty, multi-million dollar equipment, which only shows up when Arp looks through the telescope—sort of like the moons of Jupiter painted on Galileo's lens.

HALTON ARP CONTINUES THE BATTLE

This was a pattern which was to repeat itself time and time again during Arp's long career; but he did not waiver. He continued to present the evidence in a scholarly and courteous manner, amidst heaps of abuse, prevarication, and irrationality. Even when the evidence seemed incontrovertible—when his source was the mainstream body of knowledge itself—the pattern remained unchanged.

Such was the case of the Seyfert galaxies. "The American astronomer Karl Seyfert discovered this class of galaxies in the 1950's . . . by noticing that some galaxies had brilliant, sharp nuclei" (Arp 1996, 35). After the more sophisticated X-ray telescopes like the ROSAT were put into use, (in the 1990's), it was discovered that the bright galaxies with strong X-ray emissions were Seyferts. It soon became obvious that these long known and extensively catalogued sources (the brilliant nuclei), were quasars. Since Seyfert galaxies were already "on the books" and diagramed, any astronomer could now see at a glance that this whole class of active galaxies was associated with quasars (37).

When it became known that Arp intended to communicate his new findings by presenting a paper at the well known Texas Symposium on Relativistic Astrophysics; his name was removed from the roster. The same thing happened at a Tokyo conference scheduled a short time later. At the last moment, after showing enthusiastic interest in his theory, the organizers, bowing to pressure from the establishment no doubt, revoked his invitation (Arp 1996, 38). Undaunted, Arp went on to demonstrate that well observed "water masers" were associated with Seyfert quasars; and that "BL Lac objects," another known but rare, high redshift object, offer a powerful proof of a link or association between emitted quasars and new "companion galaxies" (42, 48). At this stage, Arp was beginning to receive theoretical support from physicist, Jayant Narlikar, who, with P.K. Das, predicted that quasars would reach a maximum distance from the parent galaxy of about 400 kpc (around 13,000 light years) (43).

THE THEORY

"In spite of a deliberate effort to avoid them, a large number of cases of quasars undeniably associated with much lower redshift galaxies have accumulated"

(Arp 1996, 54). From these cases follows the unavoidable conclusion that new galaxies are born within, and emerge from, older galaxies. New galaxies are not just a *part* of the older, parent galaxy, sharing the same matter. *They are made of newly created matter.* This new matter is made in the nucleus of the older galaxy in the form of a high energy plasma, which is ejected from the parent along its minor axis (The "axel" of the galactic "wheel"). The plasma is ejected in two opposite directions at speeds approaching the speed of light, but then slows down gradually as the mass of the quasars increases, and eventually forms new or companion galaxies and comes to relative rest at 400 kpc from the parent (54).

Quasars have a high redshift, not because they are moving away from us at high velocities. They may in fact be moving away—or toward us—at moderate velocities, and the Doppler Effect redshift, if there is one, is part of the calculation. But the most significant redshift by far is the *intrinsic* one. This means that *the redshift comes from the matter itself*, and the fact that it has a low mass. Again, Narlikar has formulated the theoretical explanation. Building upon the physics of Ernst Mach, Narlikar explains that the inertial mass comes from its interaction with the rest of the universe, and simply put, new mass hasn't had enough time to interact with other parts of the universe—hence its low mass. Since the mass of its new or young electrons (which emit light when exited), is low, the electromagnetic radiation is weak—hence the redshift (Arp 1996, 108). This theoretical basis may or may not be the correct one, but the fact remains that the observations, accumulated over many years, make it clear that the redshifts of quasars are intrinsic, and that the most logical explanation is that quasars are composed of new matter.

SEE NO EVIL

Although I generally don't give the mainstream too much credence—or text space—in this book, it would be instructive to review some of the kinds of responses today's representatives of this venerable science have given to Arp's well reasoned contributions. Here are a few examples:

1. A rebuttal paper was published claiming "complex orbits" to explain the "preponderance of positive companion redshifts" (Arp 1996, 74). Reminiscent of "epicycles?"
2. "There were hints that the thesis student who found the excess companion redshifts would be in big trouble" (74).
3. "No matter how many times something new has been observed, it cannot be believed until it has been observed again" (75).

4. Most astronomers are ready to suppress any and all observational evidence in order to prevent reexamination of their key assumption that redshifts can only be caused by the Doppler Effect (81).
5. Anonymous messages by referees were sent referring to "ludicrous" and "bizarre conclusions" of those who propose non-Doppler redshifts (83).
6. Arp realized that his prospects for a career in astronomy were dim since he alienated the editor of the *Astrophysical Journal*. (91). He then lost his tenured position for that reason (92). Whatever happened to academic freedom?
7. In nearby galaxies, such as the "Magellanic Clouds," and in our own "Milky Way" galaxy, excessive redshifts of bright supergiant stars were attributed to solar wind! A world expert on solar mass verifies that this is impossible (96-97).

When reviewing these and many other examples of what I call the "see no evil" syndrome, I had these passing thoughts: Although there have always been some astronomers and physicists, who maintained the proper objective, scientific open-mindedness in the face of this onslaught of this medieval Scholasticism, I realized that I hadn't previously appreciated how wedded mainstream astronomy was to the Big Bang theory and the related Expanding Universe theory. After all, they were just theories. They had no immediate impact on our lives. The layman considers them to be non-controversial, non-political issues. But to the astronomers whose livelihoods and the entire edifice of their funding is based on the current paradigm, this is a deadly serious, big business. And this brings us back to one of the central issues of this book—a problem which threatens modern science in general—namely, the public funding and government control over academic science, and therefore the power to suppress dissent.

A NEW COSMOLOGY

Something has been said on the subject of cosmology above, and more will be said when we get to the theories of Tom Van Flandern. In science, as opposed to philosophy, cosmology is really just the "big picture" of astronomy. As mentioned above, the Big Bang cosmology, states that the universe began from nothing around fifteen billion years ago. It will keep expanding until either it dies in a "final entropy" state, or until it reaches a maximum expansion point and then begins to contract until it achieves finally the nothingness point of a "singularity." Then presumably, the cycle starts all over again.

Arp's cosmology is much more pleasing to the common sense, and much more "intuitive." Although in the end, all cosmologies finish with the Unanswered and Unanswerable Questions of philosophical cosmology. Namely, where did it all come from? Is there a God? And so on, (see Chapter 2).

Arp's cosmology (Arp 1996, 225-52) is one of the continual creation of matter, and the continual birth of galaxies. This creation is not "from nothing," but from the "sub-strata" of the universe. (For more on this hypothetical concept, see Chapter 10 and Tom Van Flandern's discussion of "infinite scale"). The greatest mistake in science, writes Arp, "and one that we continually make, is to let the theory guide the model" (239). To Arp, and I agree, the empirical evidence should be, if not everything, then the final arbiter. Hence he favors what he calls the Empirical Model of the universe.

Did the universe have a beginning? Not one that we can see. But stars and galaxies definitely have beginnings, and their redshifts tell us their ages. First, a very high redshift, "super fluid" is ejected from the parent galaxy along its minor axis, as mentioned above. This super fluid has a near zero mass and travels at near the speed of light. As it travels, its mass increases and it slows gradually—eventually becoming a quasar. This quasar continues to increase in mass as the first stars of the newly born galaxy begin to form, and eventually, after a few short million years, the new galaxy takes its normal spiral form, and will in turn eject young galaxies of its own. Astronomers can see the evidence of this evolutionary process, in the alignment of galaxies along their minor axes, with little deviation over the eons. They are arranged in clusters with the oldest at the center, proceeding outward to incrementally younger, higher redshift galaxies, and then to quasars.

During this process, another interesting phenomenon is observed. Redshifts, as they change from their initial high value to their final level in mature galaxies, they change in *quantized* increments. That is to say, just as in quantum physics, the values of allowable redshifts occur only in certain discrete increments (Arp 1996, 195-223).

Thus we have a picture of a universe which is not expanding in the sense that it began from nothing, eventually expanding and dissipating into the vast nothingness of space. Instead we observer a universe which is continually renewing itself in rebirth. What is the fate of old galaxies? As Dr. Arp puts it:

We could speculate that eventually the stars will burn out and the galaxy would fade into dark ashes. But my own speculation is that as the elementary particle masses grow larger, their characteristic frequency increases and at some point they may simply vaporize back into the energy field of the universe (from a correspondence).

But Arp's Empirical Model is less concerned with speculation and more concerned with what we observe. Although there is logical certainty that galaxies, such as M31 in the Virgo Cluster, which are negatively red-shifted relative to our own Milky Way galaxy, and are therefore older, no stars in any galaxy have been dated at more that 15 billion years old (plus or minus two billion). Nevertheless, the model predicts that the universe is indefinitely, if not infinitely, old—although we have only information about our own "local" super cluster of galaxies. In Arp's model, no telescope can be sure of any observations beyond that point.

A final word will be said about the vaunted cosmic background radiation (CBR). This steady, low energy radiation comes at us from every direction in space, and has been lauded as the remnant of, and empirical evidence of the Big Bang. But we may simply be seeing, says Arp, the underlying temperature of space, which is around 2.74 kelvin (Arp 1996, 237).

We have seen, then, in these few pages, that astronomy has come a long way since the time of the Schoolmen's battle with Galileo and Copernicus. But history has a way of repeating itself.

Chapter Ten

Wrestling Superman: Tom Van Flandern's Meta Science

A PHILOSOPHER / SCIENTIST

Of all the Principals discussed in this book, Tom Van Flandern is perhaps the most philosophical. He does not merely formulate many new extraordinary, or what he calls "replacement" theories, and hypotheses. He has created a whole new system, a new philosophical approach—a new scientific method. Like other scientists treated in this work, he has started an organization to foster change in his field, and to offer a forum for new ideas. His contributions to this present work are far reaching, and indeed have already begun, (e.g., chapters 6 and 7). He has named his approach, "Meta Science."

Van Flandern's work can be characterized yet another way. Most of his major theories are the direct result of his exemplary method of deductive reasoning; but not all could be objectively considered as convincing to the same degree, and as most likely to be true. They can be divided, for convenience, into three groups or classes:

In the first class are his astronomical theories, including his theory of comets, and the Exploded Plant Hypothesis. It also includes his model of the formation of the Solar System, including the various planets and moons, his theory of gravitational spheres of influence, and much more. These are all areas in which he is definitely a world class thinker and scientist, and deserves acclaim which he is not getting. These theories, along with his basic philosophical and scientific system, are well documented and explained in terms any intelligent layman can understand, in his book, *Dark Matter Missing Planets & New Comets / Paradoxes Resolved Origins Illuminated*, (Van Flandern 1993), and further amplified in his *Meta Research Bulletin*, in his website by that name, and in numerous other writings. Again, I owe to Tom Bethel of The American Spectator, recognition for having brought Van Flandern and his Magnum Opus to my attention.

In the second class is his theory of "pushing gravity." There is also a book by that name, in which Van Flandern is a contributing author, (Edwards, et. al. 2002). Related to that, is his theory that the gravitational agents, or *gravitons*—the theoretical *cause* of gravity—are many orders of magnitude smaller than the smallest quantum particle, and propagate at a rate very much faster than the speed of light. Interesting in this regard is that unlike all previous theories of gravity, Van Flandern proposes a limited range or distance these gravitons will travel before hitting something—and hence a limited range for gravity itself. This is, of course, contrary to the Newton/Einstein theory of "universal gravitation;" the idea that all objects are attracted to every other object in the universe. If true, this one idea alone would represent a major paradigm shift in itself.

Also in this second class is his cosmology, which draws on Halton Arp's cosmology, (see chapter 9), but goes much further. This theory considers the universe to be infinite in all "five dimensions"—namely; the first three dimensions, (height, width, and depth); time, in both directions, past and future; and "scale."

Although a deductive reasoning process is necessary to any scientific theory, in this second class the deductive method has a special significance in that he uses it as a starting point to arrive at his cosmological theories. This will be explained below. Also, although these ideas can be compelling and are logical, besides running counter to contemporary theory, they also run counter to the classical strain of mainstream physics; namely, electricity and magnetism. They seem to return us to the most basic mechanical model of the universe, although Van Flandern doesn't frame it in this way. They also run counter to a more "intuitive" concept of the universe which is finite in scale but infinite in time.

In the third class belong the "Face" on Mars, (see next chapter), and other apparent artifacts of extraterrestrial intelligent design. I find the Face interesting and worthy of further exploration. But first, let us look at Van Flandern at his best.

TELL-TAIL COMETS

What are comets? These strange sights have been seen by sky watchers for as long as memory. Every so often, one is seen moving across the night sky—a brightly lit object, with a long glowing "tail." Very occasionally, one may even hit the Earth. For a non-technical, almost poetical description of comets from a thoroughly mainstream point of view, see Carl Sagan's classic, *Cosmos* (Sagan 1980).These are not "shooting stars" or meteorites burning up in the Earth's atmosphere, to be quickly extinguished. These are comets, long lasting visual events with bright, glowing tails.

Centuries ago, they were discovered, by Halley and other astronomers to be rapidly moving objects in great elliptical orbits around the Sun. They are seen entering the Solar System when their tails first light up in the region of Saturn. They plummet toward the Sun and then around it in orbits which can have periods of hundreds, thousands or even millions of years. Almost all of that time, of course, is spent beyond the planets of the Solar System in the vast reaches of interstellar space. It is now known that "new comets," i.e., comets which have never before approached the Sun, can have orbital periods of over three million years! (Van Flandern 1993, 187, 405). But what they are, and where they come from, has long been a mystery.

Astronomers have hypothesized that comets are like "dirty snowballs," with nuclei of up to several kilometers in diameter, composed of frozen volatiles and dust. As these volatiles are heated by the Sun, the vaporizing gases reflect the Sun's rays and are seen by us as a long, glowing, streaming tail. These "dirty snowball" comets are theorized to have been formed in a great nebula outside our Solar System, now called "Oort's cloud of comets," which is supposed to exist at 1,000 times the distance of the outermost planet, Pluto. But there is scant proof of all this. As usual, the mainstream won't relinquish any of its Canon Law; its theories persist and dominate. Apparently careers and "prestige" take precedence over facts and evidence.

For one thing, the point in the Solar System, near Saturn's orbit, where the comets' tails "light up" is too far from the Sun for the volatiles to vaporize from solar heat. Next, photographic images of comet nuclei show them to be solid objects like asteroids, and not merely frozen volatiles—not "dirty snowballs" of gas and dust. Moreover, other objects and debris—satellites, dust particles, and volatiles—are seen to be gravitationally bound to comet nuclei, and in orbit around them! All this is in opposition to the conventional theory. And in addition—there is no "Oort's cloud" to be found. Also not surprisingly, a more reasonable, more viable theory, one that fits the facts much better, has been around for a long time. It took Tom Van Flandern to connect the dots.

THE EXPLODED PLANET HYPOTHESIS—HISTORY

In 1772, the astronomer Titius discovered the fact that each of the six known planets was roughly twice the distance of the previous one from the Sun. Except for one—there was no planet in the space between Mars and Jupiter. In

1801 the astronomer Piazzi discovered a "tiny planet" in the exact place where Titius predicted there ought to be one. Soon after, another "small planet" was discovered nearby (Van Flandern 1993, 157). From this, it didn't take Heinrich Olbers long to figure out what had happened. The large planet which had once occupied that position had exploded! He rightly predicted that many other asteroids and meteors would be found in this region, which is now known as the Main Asteroid Belt (158). But the jealous mainstream is forever vigilant in its prerogatives. The famous astronomer Laplace soon attacked Olbers' theory because it conflicted with his own "nebular theory" for the origin of comets (158), a theory which was similar to today's "Oort's cloud of comets." Olbers' reasonable theory thereby fell into disfavor; a hypothesis for "crackpots" and for those not concerned about getting ahead in the world.

Fast forward 200 years: In the year 1972, Michael Ovenden revived Titius's and Olbers' theory, and hypothesized that a giant planet the size of Saturn, had exploded, scattering debris throughout the Solar System and outside of it. However as Van Flandern points out (159), in the first 100,000 years or so, after such an event, Jupiter and the other planets would have swept up most of the fragments which were in irregular orbits, leaving only the nearly circular orbits of the Asteroid Belt which couldn't collide with the planets; and also leaving debris which would have been ejected from the Solar System as fast moving fragments of the explosion—but not far enough to escape the Sun's gravitational pull completely. Eventually, up to more than three millions of years later, these last would return as comets (187).

In 1978, Van Flandern introduced a theory "that the comets originated in the energetic breakup of a body orbiting the Sun in or near the present location of the Main Asteroid Belt in the relatively recent past" (Van Flandern 1993, 185). This would be Ovenden's giant planet, but Van Flandern had good evidence, and a logical "*a priori*" hypothesis. It had to do with the fact that asteroids, and even probably comets, have satellites.

SPHERES OF INFLUENCE

An excellent example of Van Flandern's method of reasoning is connected with the principle of "spheres of influence." As an object moves away from a dominant gravitational field, its own gravity begins to take over. As Cavendish had demonstrated in the eighteenth century, even small objects have gravitational attraction for each other, and this can be seen once the Earth's gravity is neutralized (Shamos 1987). During the Apollo missions to the Moon, astronauts found that the garbage they had jettisoned continued to follow them all the way to the Moon—orbiting their space craft (Van Flandern 1993, 140). Thus Cavendish's lesson was relearned in a new an interesting way. The spacecraft had its own gravity—once it was freed somewhat from the Earth's dominant field!

In the case of a comet, the further it moves away from the Sun, the larger becomes its sphere of influence, meaning the area in which it can gravitationally hold other objects. As it moves toward the Sun, this area diminishes, the extent of which depends on how close it ultimately gets to the Sun (Van Flandern 1993, 139, 144, 206). Now here's where Van Flandern's bold and compelling type of reasoning comes in. As astronomers began to see and to photograph, dust, debris, objects, and volatiles, attached to and apparently gravitationally bound to comets, instead of concocting far fetched theories about "frozen snowballs" and comets "splitting," they could, like Van Flandern, have found a much more logical explanation.

How could so much material be "attached" to these comets? It was known to astronomers how extremely difficult, for example, it is for even much larger objects like planets and moons, to gravitationally "capture" objects outside their spheres of influence. So how could relatively small comets do it? The answer was found, by Van Flandern, in the Exploded Planet Hypothesis (EPH), which had been rejected by mainstream science. When Ovenden's planet had exploded, it sent debris careening in all directions into space. As this debris moved out into space, farther away from the gravitational attraction of the Sun and planets, and as the spheres of influence increased around such debris, they became gravitationally bound to whatever dominant (largest) fragment, meteor or asteroid, was traveling in the vicinity, in the same direction and at similar velocities. Those objects thrown the farthest became the comets (Van Flandern 1993, 211–12). Simple right? But to accept the obvious logic of this, you have to accept the EPH; oh, but there's the rub; the mainstream's canon law gets in the way again.

To follow this logic a little further, as a comet approaches the planets and the Sun again, after its long elliptical orbit into interstellar space, the sphere of influence begins to shrink, forcing dust and volatiles to be released from gravitational attraction to the nucleus, they then vaporize and light up the night sky in the form of the comet's tail! So Van Flandern showed what should happen logically, according to known laws of physics; he came up with his "extraordinary hypothesis," or what we call, his Apocryphal theory (Van Flandern 1993, 355). To follow the logic further: "novas" were long known by astronomers to be dying stars in their last explosions of light and energy—going out with a bang, so to speak. But many of them were *previously unseen* "stars" (156). Isn't it far more likely that they were really planets exploding? I mean, *if they were stars, they would have been seen.*—And further: comet orbital trajectories can be traced back in space and time. Their point of origin? It was Ovenden's planet between Mars and Jupiter.—And further: there are telltale "black axioms" strewn all over the Solar System, black carbonaceous deposits on moons with slow rotations, which were blackened on one side by the blast wave of the explosion event (281, 290). Mars itself, which was probably originally a moon of an exploded planet, (see below), shows clear evidence of massive debris covering one hemisphere (427). This is but a small sample of Van Flandern's logic and evidence.

THE REAL SOLAR SYSTEM

Van Flandern favors the well known "accretion model" for the formation of stars, including our own Sun; but he has added new insights, many of which, we say again, are rejected by the mainstream. In the standard model, stars condense out of huge clouds, or nebulas of primeval gas and dust. In our Solar System, after the Sun condensed, the planets are said to have condensed from the remaining nebula surrounding the Sun. But this theory is vague and leaves many questions unanswered, such as, how do the fundamental particles or "baryons" coalesce and "stick together" in the first place? Van Flandern offers a more logical and elegant model. This description will necessarily be very brief and simplified.

Through electromagnetic and gravitational forces, huge amounts of matter accrete and coalesce to form a star, or the Sun in our example. As it condenses and heats up, the heavier matter sinks to the center, thus causing it to spin rapidly— as a ballerina spins faster when she pulls in her arms. Finally it reaches "overspin," meaning that speed of rotation is sufficient for surface layers to be thrown off by centrifugal force and go into orbit around the Sun. This transfers some of the Sun's angular momentum, and thus spin, to the thrown-off matter, which itself coalesces under its own gravity to form a planet. The Sun then slows its spin to below the overspin state, and again continues to condense, causing it to overspin once more, and throw off another spinning "planet." This process is repeated a number of times until the Sun finally achieves a stable condition and no longer condenses and overspins. Now we look at the planets. As they accrete and condense from the thrown-off matter just described, they in turn heat up and reach overspin. And they in turn throw off matter to form their own moons. The larger planets, such as Jupiter, formed several moons in this way; the smaller planets, such as the Earth, formed only one (Van Flandern 1993, 328–32). In the case of some moons, like our own, gravitational and tidal forces on them are so great that they eventually stop spinning completely. This simple, logical model is of course rejected by the mainstream, in favor of less defensible models.

Van Flandern then extrapolates from this model, and from extensive observational evidence, to formulate a theory of what the original Solar System might have looked like (Van Flandern 1993, 332–36). In this model, Mercury, which is only four times larger than our own Moon, was originally the moon of Venus. There was originally a "Planet K," (named for Krypton, the mythological planet of Superman), which had Mars for its moon. There was originally a "Planet V" or fifth planet, which is at the location of today's Main Asteroid Belt. Pluto was originally a moon of Neptune. And then there is "Planet X," a large and as yet undiscovered outermost planet. This model may seem far reaching and extravagant at first glance, but after reading Van Flandern's reasoning process, based on his extraordinary knowledge of planetary dynamics, and after examining the ever increasing evidence, the reader will agree that the model is compelling.

But when objections came in from various quarters, some of it valid, and along with it new observational evidence, Van Flandern latter modified both his Exploded Planet hypothesis (EPH), (Van Flandern 1993, 406–09), and his model of the Solar System (444–54). For example, in his Solar System revision, he explains why planets form in pairs (448). Regarding the EPH, multiple catastrophic asteroid impacts at the K/T Boundary, the date which marks the extinction of the dinosaurs, and other evidence, strongly suggest that the explosion of the original fifth planet occurred 65 million years ago (Ma), instead of 3.2 Ma as originally hypothesized. But the existence of the comets still indicates that something, probably a smaller object, exploded 3.2 Ma. That's a lot of explosions, a scary prospect if true; but that's where the facts and logic lead.

A WEALTH OF LOGIC AND FACTS

The above only scratches the surface of what Van Flandern has given us. Other examples of his science which I could have used are; the difference between captured moons and natural moons; the dynamics of gravitational capture—and escape; Jupiter's Great Red Spot, which is really a moon-sized object floating in Jupiter's heavy atmosphere, and not a "global storm;" Saturn's Rings, which are remnants of an exploded planet; the dynamics of tidal forces; and much more.

What I did relate to the reader, is an example of cutting edge science at its best, science with little or no political ramifications, yet rejected by the mainstream just the same. What difference could it possibly make how the Moon was formed, so long as we know that it will last many billions of years more? Why not be open to the facts and interested in the truth? This is all the more reason to be concerned about the present state of affairs in science, and to try to do something about it.

ZENO

If I am in less than complete agreement with Van Flandern's theories in the following, I do so in the spirit of free inquiry, and in the interest of science—though admitting cautiously as I proceed, that disagreeing with him is as hard as trying to wrestle Superman. Nevertheless, as discussed previously, mistaken philosophical formulations can lead to mistakes in science.

The basic idea of Zeno's paradox(es) is that a man, or a turtle, can never get from "here to there" because every time he passes the half-way point, he has to pass another half-way point of the remaining distance. The problem is that there are an infinite number of (theoretical) half-way points, and therefore he will never get "there"—even if "there" is only across the road, or across one inch. But since he *does* get there, we have a "paradox," or apparent contradiction.

Van Flandern's contention is that this paradox is significant because it represents a fundamental description of reality (Van Flandern 1993, 6). As was mentioned, Van Flandern postulates a universe which is infinite in five dimensions, and that "scale" is one of the dimensions. This causes me to "raise an eyebrow" (like Mr. Spock). Here's why. On the small scale, the smallest known particle can be an infinitely large universe on some smaller scale. On the large scale, a "wall of galaxies," trillions upon trillions of them, might be a mere light wave in the quantum world of some larger scale. That's it, except to say that you can repeat that statement an infinite number of times and it would always be true!

Van Flandern is deadly serious about this, because he thinks that it is the only way to resolve certain logical/mathematical dilemmas, among which Zeno gives us our first hint. I'll use an example he gives in *Dark Matter* (1993, 7–9) in order to illustrate my objection. In the example, points X and Y are separated from each other by the "smallest possible distance" (in the universe). If you then take point Z and place it also at this smallest possible distance, but at a slightly different angle from say X, then the distance between Y and Z will be *smaller* than the "smallest possible," which is of course illogical—and paradoxical. Van Flandern says that the only way around this dilemma is to admit that there is no such thing as a "smallest possible distance." By doing so, you arrive at an "infinitely small scale."

My objection to this hypothetical construct is that you can't use something as a starting point that you don't know for certain. In philosophy, this amounts to assuming as true that which you are trying to prove. For example, you can not automatically assume that you can employ any "point" as small as you please, in order to illustrate your construct. If particles X and Y are really at the smallest possible distance from each other, then two things are necessary —two prior requirements: first they must be the smallest possible particles (in the universe), and second they must be touching each other. If this is the case then particle Z, off in a different direction will also be touching, and also at the smallest possible distance—thus resolving the paradox. In contrast to this, in Van Flandern's model, X, Y, and Z are assumed to be *indefinitely small points*—thus he arbitrarily assumes to be true precisely that which is necessary to prove his hypothesis. It is the original problem of the syllogism coming back to haunt us, namely that the major premise must be factually correct for the syllogism to be true—in fact.

In his chapter on scientific method (1993, 347–62), Van Flandern makes a very good case for better use of the deductive method in formulating scientific models. His theory of comets is a good example of how he does this. But in the case of infinite scales, that method may be inappropriate. In logic, argument from generals to particulars (deduction), and vice versa, (induction), must *both* be used. But what Van Flandern has done in his cosmological model, is to rely solely on deduction to form his hypothesis. To put it simply: it may be that the universe is infinitely small, but there is no proof of this. Zeno just can not serve as proof. Neither can mathematical constructs when their referents are purely hypothetical.

As for examples of the type of logical flaw employed in Zeno's paradoxes themselves, they are as follows: they overlook the fact that degrees of smallness vary directly with the time it takes to traverse them. Given a constant speed, the smaller the distance, the less the time; infinitely small distance, infinitely small time; no distance, no time. It also overlooks the fact that "infinitely small point" is a mathematical construct with no known referent in reality. It therefore amounts to using an arbitrary or imaginary construct as the basis for proof.

Infinities or no infinities, the universe is whatever it is. If there is a certain object that is the smallest—whatever it is, then that's it. On the macro or cosmic scale, if there is a point where there are no more galaxies, then that's it. We need observation and induction to find out what these limits are, as best we can. We need deduction to make sense out of what we find out. To make sense out of the particulars we observe and think about inductively, we construct logical deductive models to best explain them.

That said, scales much smaller than the quantum level may nevertheless be logical and necessary to explain gravity. They may be the best way to explain the phenomena.

TWO SUB-ATOMIC SCALES

There is a lot of good deductive proof which makes the idea of "pushing gravity" compelling. But there are still some good logical objections. Although Van Flandern deals with the idea of pushing gravity in *Dark Matter*, (1993, 27–53), and in several articles and papers, (e.g. June 2003), his article in the book *Pushing Gravity*, (2002, 93–121), is quite comprehensive and will be my primary reference for this subject.

Related to the idea of infinite scale, there are two hypothetical scales fundamental to this theory. There is the "light carrying medium," (LCM, or "elysium," as Van Flandern calls it), and there is the "graviton medium," several orders of magnitude smaller than the elysium, but made up of particles which travel some 20 billion times faster than light, (ftl) (Van Flandern 2002, 106). The "elysium" is the "ether" of 19th century physics, (see chapter 6). Elysium particles are several orders of magnitude smaller than the quantum scale. If all this is getting confusing, the following is a simplified and brief synopsis of the idea of "pushing gravity."

PUSHING GRAVITY

Have you ever wondered what causes gravity? Some of the greatest minds in history have pondered this question—and then pretty much given up on it. Newton, and later Einstein to a more exacting degree, gave us mathematical models of gravity, which, on our scale, hold true to this day. Except that they say nothing about the mechanism which actually causes gravity. Einstein's theory, General Relativity (GR), attributes the cause to the "fabric of space," (see Chapter 6). But as Van Flandern points out, (Van Flandern 2002, 94), Einstein's "rubber sheet analogy" presumes *real gravity* underneath the "fabric," which causes planets to "sink" down into the "gravity wells" in the sheet. It therefore explains nothing about the real cause of gravity.

In the mid 18th century, G.L. LeSage proposed a mechanical theory of gravity whereby tiny particles in space move about in all directions and at very high speeds, causing equal force on all sides of any object or planet they make contact with. But the space between any two objects has less of these particles or "gravitons" than the surrounding space, because some of the gravitons have already been absorbed as they passed through the object. This dearth of gravitons between objects causes a kind of low-pressure area allowing the gravitons in the outlying areas to push the objects together—thus causing gravity.

LeSage's theory has been revived and then rejected by many famous scientists over the years, and is presently undergoing its most recent revival. If such particles exist, there must be a way to detect them. One method, attempted by the physicist Q. Majorana, early in the 20th century, was to test the effect that gravitational shielding has on the absorption of gravitons and hence the weight of objects. Through elaborate scaling experiments carried out in a very careful and meticulous manner, he obtained some seemingly positive results (Martins 2002, 219–38). But these results, along with more recent shielding experiments using satellites in Earth orbit, have not been conclusive. They have not yet been demonstrated to the satisfaction of other scientists.

What the theory of pushing gravity does have, especially in its modern versions, perhaps best exemplified in Van Flandern's model, is its compelling logic. It posits a physical cause to gravity as opposed to, for example, Newton's instant action at a distance, (IAAD), which, incidentally Newton was never happy with; and Einstein's "fabric of space," which simply "begs the question" of what causes gravity.

Van Flandern points out that IAAD can be better explained by positing gravitons which move at many times the speed of light. A slower propagation speed, e.g., the speed of light would cause a delay or *aberration* of the signal carrying the gravitational force between, say, the Sun and the Earth. This would cause Earth's orbit to be unstable.

There are many implications and possible new properties for this model of gravity. In one example, if gravitons travel in a straight line until they encounter an object, or "matter ingredient" and are absorbed, there should be a maximum range for gravity. This is a point beyond which the "shading effect" between two objects diminishes to the vanishing point and there is no gravity between the objects. Van Flandern calculates this distance to be around 6,000 light years, which is less than the size of our galaxy. This means that galaxies and clusters of galaxies are not bound gravitationally, beyond local groups of stars within galaxies, and it explains why stars at the tips of the spiral arms of galaxies appear to be moving out into intergalactic space. This is indeed a revolutionary idea, but Van Flandern says that it can be tested through observations in our solar system where a progressively decreasing gravitational constant should be measured for reference frames further from the Sun (Van Flandern 2002, 107–08).

CONCLUSION AND OBJECTIONS

On the face of it, some variation of LeSage's theory seems plausible as a physical or mechanical explanation for what actually causes gravity. But I have certain intuitive objections, or at least questions which raise doubt in my mind. I'm sure Van Flandern can answer all of them—still we will pose them.

1. After at least a century of trying, there is still no physical evidence of such gravitons, meaning that no-one with the possible exception of Majorana has been able to manipulate them in any convincing way. Van Flandern notes in this connection that such evidence of "gravitational shielding" has possibly been detected observationally in the Lageos satellites (Van Flandern, December 2003).

2. Hypothesizing particles many orders of magnitude smaller and faster than light is reasonable if that proves to be the best explanation for the phenomenon. But, as discussed above, one need not pose infinite smallness to buttress this hypothesis.

3. If gravity seemingly acts instantaneously, there are other possible explanations besides super-fast moving particles. Beckmann has offered one tentative explanation to the problem that orbits would decay if gravity propagates at the speed of light, and if gravitational aberration was factored in, in the same way as light-time delay. What he proposed is a corollary of the "discrete permissible orbits of electrons," similar to that established by quantum physics. He proposed that at certain discrete distances, namely those of the Titius-Bode series, the "aberrational component" is neutralized. (Beckmann 1987, 179). Other contributors of the book, *Pushing Gravity*, have proposed similar ideas (Jaakkola 2002, 155–166).

   Another possibility is that gravitons do exist, but they travel at slower speeds, or are themselves a form of (Machian) electromagnetic force, but they do exert a "pushing" force on objects. The reason no aberration from these slower moving gravitons is observed, is perhaps because the incidence of force on all objects is *simultaneous*. Gravitons do not need to propagate at super speeds if, coming from different directions, they act on the attracting bodies, *at the same time*. The following is Dr. Van Flandern's reply to this objection:

The Sun and Earth are a finite distance apart. So if either one causes an effect, such as a graviton shadow, it takes a finite time for that shadow to reach the other. The two bodies also have a relative motion. So when the propagation-delayed shadow reaches the other body, it will appear to come from the retarded position of the first body, not its instantaneous position.

You can think of a single unbalanced graviton instead of a whole shadow of missing gravitons. Compare the two cases, one where the Earth is bathed in a uniform, continuous rain of gravitons from all directions and feels no net force, and the other where a distant, moving body blocks a single graviton. That missing graviton will cause the Earth to feel a small net force toward where the moving body was when it blocked the graviton, but only after the graviton would have reached Earth if it had not been blocked. So the propagation time to Earth is indeed relevant. (from a correspondence):

4. Another intuitive problem I have with the idea of pushing gravity has to do with electromagnetism itself. It is hard to imagine that a magnet attracting or repelling another magnet has anything to do with mechanical "pushing gravitons." Faraday's "field" is a quite elegant explanation. On the other hand, the "electrical forces" we were weaned on, as with Newtonian gravity, do not really explain fundamental causes in the way that pushing gravity does.

There is always a risk in challenging a specialist in any field armed only with the weapons of mere logic. The problem is that most of the weight of any argument is in the details. As this manuscript was in its final stages of revision before going to press, Dr. Van Flandern has "taken the plunge" in formulating a deductive model for electromagnetic phenomena at the quantum level. While a detailed review of his new theory is beyond the scope of these concluding remarks, a hint of what this fertile mind has come up with will be presented here.

Quantum particles such as protons and electrons have unique properties which allow them to behave in ways unlike "ordinary matter" (i.e., matter at our scale), yet still to conform to pushing gravity theory. For one thing, ordinary matter is mostly "empty space," meaning that there are immense spaces between the components which make it up, especially at the quantum level. Therefore the much smaller gravitons which bombard it, pass through for the most part unhindered, making only enough contact with "matter ingredients" to cause gravity. But at the quantum level, the much denser protons, for example, do not allow gravitons to penetrate to their centers. One result is that the "elysium atmosphere" around each proton is so dense that it causes protons to bounce off or repel each other with a force much greater than the gravity which attracts them. But electrons have, instead of this elysium field or atmosphere, an "elysium hole," causing them to attract protons to fill the void. There is much more but that's the idea (Van Flandern, December 2003).

We will continue with Van Flandern's theories in the next chapter.

Chapter Eleven

The Face:
A Fact that could Change Everything

EVIDENCE THAT WON'T GO AWAY

If there is a face on Mars, sculptured by intelligent beings out of metal or rock, then that fact—if it is one—changes everything. What do I mean by "that fact changes everything?" Only that if we had proof of such a thing, it would mean that "we are not alone," that there is other intelligent life in the universe beside us. If we could be sure of such proof, that would be a story without rival in human history. That hackneyed phrase, "we are not alone," has been so overused by UFO enthusiasts, that it may be viewed with skepticism. I suggest that in this case it may be appropriate.

Such a face has been photographed by the National Aeronautics and Space Administration (NASA), at first accidentally, when it was discovered in a series of images taken by the Viking spacecraft while surveying and mapping the topography of Mars in 1976. Subsequently, it was reluctantly photographed two more times; once in 1998, and once more on April 8th, 2001. The initial accidental image was found in a batch of NASA images by engineers, Vincent DiPietro and Gregory Molenaar, in 1979, (Hoagland 1996). The idea that it might be an artificial structure was quickly ridiculed.

The image might have been consigned to oblivion were it not for interested astronomers like retired U. S. Naval Observatory astronomer, Tom Van Flandern and his organization, Meta Research, Inc., (see chapter 10), in bringing public pressure to bear on NASA and Jet Propulsion Laboratories (JPL), the contractor carrying out the unmanned Mars exploration project. It was only through such efforts that NASA agreed to photograph the face again.

After the most recent (2001) image was taken, press releases reported that the government's previous opinion was now confirmed. The so-called face was just a "pile of rocks." It was merely a natural formation which happened to vaguely resemble a human face as geological formations and clouds sometimes do.

But those astronomers who think otherwise, namely that the face *is* what it appears to be, appeal to the scientific method to help evaluate the evidence. A careful analysis of the various images taken, especially the most recent one which is the best and the clearest, is under way (Van Flandern, June 2001). Also being studied are certain other curious structures imaged during the surveys of Mars. This study, if free and open, will hopefully decide the issue one way or another.

Meanwhile, government spokespeople for NASA and JPL have resorted to ridicule and ad hominem attack. Only "crackpots" and "true believers" see "faces" on Mars. This is the stuff of science fiction, they say, like UFOs and "ancient astronauts." And the fact is that many people, possessing good common sense, would agree with that assessment. It's not that they trust the government implicitly to tell the truth or to do the right thing. It's just that, on the face of it, it seems like such a far fetched notion—men from Mars? There are all sorts of reasons why the very idea might seem ridiculous. For one thing, man evolved only here on Earth. For another, Mars does not have an atmosphere capable of sustaining life as we know it. It has no liquid water, no vegetation, and extremely cold temperatures. Add to this that human civilization is only a few thousand years old. Prior to that, men lived in a primitive state, capable of using only stone implements. That's what the archeological record shows; that's what science now believes.

But these "interested astronomers" claim that the face is an artificial structure, and possibly over *three million years old*! As mentioned, one of the most credible of these astronomers is Van Flandern, who concludes by a series of inferences, that the face, if artificial, was built prior to Mars' most recent pole shift, which would place the date at prior to 3.2 million years ago (Van Flandern 1993, 417–41).

SPECULATION

If true, what does this imply? Let me speculate, if only to play devil's advocate. It means that an alien civilization may have existed, either on Mars itself, or elsewhere. Members of this civilization were either natives, or visitors, or colonizers of Mars at the time the face was constructed. It means that this civilization is comprised of a race which resembles, fantastically—*us*. This implies in turn, that either separate, independent civilizations can exist on different worlds which have inhabitants who look alike; or that the Mars civilization (or visitors), predates ours and that they are somehow involved in our evolution or development.

This often ridiculed line of reasoning has been raised in such quasi-scientific books as Erich Von Daniken's, *Chariots of the Gods*, and Zecharia Sitchin's, *The Twelfth Planet*. But this new evidence, if true, gives powerful new impetus to such reasoning. Perhaps a superior race (or races) from somewhere in the universe has long intervened in the development of life on Earth. They may even have "seeded" the Earth with the stuff of life (DNA) in the first place, allowing evolution to take its course but also occasionally intervening and redirecting it. This is not such a far fetched idea, once you get used to it.

For one thing, many scientists now concede that the galaxy may very well be teaming with planets capable of supporting intelligent life. For another, owing to its complexity, some scientists, such as Michael Behe, in his book *Darwin's Black Box*, have hypothesized that DNA, and life at the molecular level in general, is far too complex to have evolved under the classical Darwinian model–especially not in the time frame allowed (see chapters 2 and 14). These scientists have argued for "design" and against evolution, as the more plausible hypothesis at this level. This "design model" does not necessarily mean "divine creation," and it does not contradict the theory of evolution by natural selection; it is merely concerned with the question of how life at the molecular level got started and how it spreads throughout the universe.

So these space faring "designers," may have had a hand in the development of the planet Earth from the very beginning, at first seeding it, and later manipulating evolution and still later, us, just as *we* manipulate cultured plants and domesticated animals, and are now taking our first halting steps in genetic engineering and space exploration.

This is all, of course, pure conjecture and speculation, but of the kind which evidence such as the Face allows us to proffer.

BACK TO THE FACTS

But I am a person who is committed by disposition to understanding only by a process of cause and effect, or in other words, by means of reason, and I accept only definitive proof. Such theories as described above, although intriguing, do not provide that proof. I therefore am not concerned primarily with broad speculation, and not convinced by it, though some theorizing of this kind is needed in order to have a grasp of the possibilities involved, and to show that they are sometimes just as plausible as the traditional models. I am more interested however, in events that can clearly and reasonably be

Figure 1.   From raw data of image taken on April 8, 2001 by the *Mars Global Surveyor*.
It was released to the public by NASA in cooperation with Jet Propulsion Laboratories,
(JPL), and Malin Space Science Systems, (MSSS), on May 24, 2001. (Meta Research web-
site; http://metaresearch.org/).

shown to have happened, meaning *real* events which have left some kind of
a *physical* record for us to examine. Such is the case we have before us at
present.

    If you the reader were to look at the carvings on Mount Rushmore, or im-
ages of same, you, being sound of mind, would have no trouble seeing that

they are products of art, constructed by men to look like human faces. The conceptual process that would draw you to that conclusion might be intuitive or unconscious, but you would come to it just the same, and you would be absolutely convinced by it. It is not difficult to think succinctly and lucidly, especially about matters as familiar as a human face.

That is exactly the point, since the human face is so familiar. There are thousands of bits of information in the human mind—perhaps more—which go into the recognition of a face. Not only can we recognize a face as such, but also innumerable individual faces. Parts of a face can be covered up, still leaving it clearly recognizable. It is true that in borderline cases, when a face is so distorted or obscured, that we require a formal scientific process to determine if the thing we are looking at really is a face or a human face. But I think the Face on Mars goes far beyond that point. It is easily recognizable as a face and it is probably human or human-like.

Look at the Face and judge for yourself. (See figures 1, 2, and 3.) Images shown below are taken from the raw data of the high resolution NASA image

Figure 2.   Cropped and enlarged from Figure 1 by author.

Figure 3.   Cropped and enlarged from Figure 1 by author.

taken April 8th, 2001, available at the Meta Research web site. They have been cropped, and adjusted for brightness and contrast but otherwise unaltered. At first glance we see a mesa-like structure, facing skyward, around ½ mile in length which is in the shape of a human face with a frame or enclosure around it. This frame is symmetrical both on the inside and outside borders. Not perfectly symmetrical, there has been some damage, caused apparently by a meteor, which also obscures some of the face's features. The face is three dimensional when viewed from an oblique angle as in the image taken in 1998 (not shown). It has a symmetrical "head piece, " or smooth, combed down hair. It has a well defined right eye, complete with clear outlines or borders, an iris, "crow's feet," at least three lines under the eye, an eyebrow, and an eye socket. It has a nose, very wide and apparently damaged or different from most human noses; but with a bridge and two discernable nostrils. It has half of an upturned upper lip, a mouth opening, and possible front teeth.

Most of the left side of the face (the face's left, but right as we look at it), is covered with what looks like melted and re-hardened lava or metal, obscuring most of the features on that side of the face. But the left side of the forehead or headpiece, and hairline is still clear, and symmetrical, although slightly damaged, as is the left eyebrow. Part of the left eye socket and part of the left eyeball or iris is still visible although predominantly covered by the melt. The crater from the probable meteor event is on the left side of the jaw or chin, but the rest of the chin appears to have a beard!

A possible fact that I noticed some time after preparing the draft of this article for inclusion in *Apocryphal Science*, is that when looking at the 1998 image, the probable location of the damaged mouth comes into focus, making it more probable, and logical, that what we previously thought was the mouth, may be a mustache! Think of it, a sculpture of a "god" or a king, *with a beard and a mustache*, not unusual at all.

Even if one or two of these descriptions don't prove to be exactly correct, what's clear is that this is no pile of rocks. There are far too many "coincidences" for that to be the case. For a more scientific analysis of the above description see; (Van Flandern 2001).

Van Flandern's group adds other interesting "coincidences." There is astronomical evidence that the Face happens to be located on the former equator of Mars with an upright North / South orientation. There are other interesting possible artifacts such as "pyramids," a "fortress", possible ruins of a "city," partially covered "tubes" which might be conduits or tunnels of some type, possible "vegetation," and several outlines of "animals," and "people." These are all found in the same area, the Cydonia region, where the Face is located (see Meta Research website).

Another well known figure, Richard Hoagland, has popularized this issue for some years, even speaking at the United Nations in 1992, writing a book, and producing a video tape of that event. To his credit, he garnered much interest in the subject, and must be credited as being its first major popularizer. As we have mentioned, the fact that further photographs were taken attests to his efforts, in addition to those of Van Flandern and a few others. But having only the low resolution 1976 photo to work with, Hoagland resorted to what many believe is an arcane type of pseudo-science, associating for example, the Cydonia region with the pyramid region in Egypt, and the Face on Mars with the Sphinx, and intimating that the Face is somehow a door to another dimension! (Hoagland 1996). Like Von Daniken before him, he may ultimately be proven right, but that still doesn't qualify his efforts as good science.

Be that as it may, facts are facts. The Face is staring us in the face in the form of the high resolution image now in our possession. That can not easily be ignored.

THE A PRIORI METHOD

I will now address Van Flandern's method of quantifying his position that the Face must be artificial, using what he calls the *a priori* method, mentioned in the previous chapter (Van Flandern, June 2001 and December 2003). Although the method is valid, (see below), it seems to me that although predictions can be made using this method, such predictions can't be precisely quantified. But my main objection however, is not to the *method*, but to its *name*. The *a priori* concept was used by Kant to designate an idea which exists in the mind *prior* to any sensory perception or evidence of it, and as such it is linked inseparably to Kantian subjectivism and, in my opinion, irrationalism; and to the notion of "intrinsic knowledge," all of which have been challenged by Objectivist philosophy. However, although the name is unfortunate, as used here the term is more narrowly defined, and is essentially a valid concept, as I will explain.

As mentioned above, if we saw an image of Mt. Rushmore, we would know intuitively that it was an artificial as opposed to a natural formation. But let's imagine some distant future date when Mt Rushmore is no longer known to us. At first we see only a vague, poorly defined image of the mount, and for fun, let's say that Ronald Reagan was one of the presidents carved on it. And suppose that the carving was in ruins, with noses broken off, and the like. The question is; would there be some way to be certain that this vague image was artificial and not natural?

The *a priori* method says that there is. It works like this: suppose that in the first image we see, or as seen from a great distance, very few of the details can be ascertained. We can see only the general shape of the faces but not the details. For example, we can not see any of the presidents' eyes, but only shadows where the eye sockets should be. We can not see Lincoln's beard or TR's pince-nez, and although Reagan seems to be holding something, we can't tell what. The *a priori* method predicts that it is highly improbable, or even virtually impossible, for predictable details which we have not yet seen, to actually exist upon closer examination—unless the carving is man-made that is.

But as we approach the carving and take a better look, or view a higher resolution image, we *do* see the "whites of their eyes." We do see the beard, the glasses, and Washington's wig. And that thing Reagan is holding? We now see written on it in bold letters, "Wall Street Journal," and under that the subheading, "What's News?" According to the *a priori* method there can now be no doubt that what we see is man-made, not because these shapes could never have formed naturally, because given an infinite amount of time, they could; but because we could never have predicted known artificial forms before the fact—or *a priori*, unless the object was indeed artificial.

Such is the evidence concerning the face at Cydonia on Mars. None of the details described above were visible in the original 1976 image which revealed only the general figure of a "face" with all of the details hidden in shadows or not discernible due to the poor resolution of that first image. The probability that these details would be revealed in later, higher resolution images would be negligible—unless the face was the real McCoy that is.

Chapter Twelve

Over the Line:
Exploring the Borders of Science

ZECHARIA SITCHIN'S TWELFTH PLANET

People can have many kinds of motivations for believing or disbelieving a book. Does it stray too far from what we think to be true? That's always a danger. Whenever we go off the beaten path, we take a risk. We need to make sure that we have good reasons and evidence to support the positions we have taken. Either that or we must make it clear that the book is not serious, that it is "New Age" literature, or that we are "only kidding." Well, I'm not kidding, I'm talking about science here, and that is what I have tried to do throughout this book. I'll have to let the reader decide if I got it right, or at least if the theories discussed are reasonable.

Zecharia Sitchin, the next Principal we will examine, stretches the borders of reasonableness and credibility to the limit, and according to the accepted rules of science—no doubt goes over the line. But for reasons I hope will become clear, I have decided to include him in my survey of Apocryphal scientists.

First in Sitchin's "Earth Chronicles" series, his book *The 12th Planet* (Sitchin 1978), is a history of human origins which is a curious synthesis of religion and science, but not exactly good scholarship. Sitchin rarely cites sources, though he does provide a bibliography. But more important, the artifacts and historical records he studies do not really prove his case. They merely prove that such stories existed, as myths or legends perhaps, but not as real historical events. Science needs corroborating evidence.

Actually, believers in frequent UFO sightings and alien abductions might be Sitchin's most likely audience—except that his work is more scholarly than the average UFO book, and it takes a certain level of education and intelligence to get much out of it. This is not to imply that I think UFOs don't exist, only that the method usually used to deal with the subject is unscientific. Be that as it may, Sitchin's research is interesting in many ways and deserves some consideration.

My impression in a nutshell is this: it may be true—at least some of it—because some of it has a certain compelling logic to it. Though on the other hand there are gaps and holes. For example, if as Sitchin claims, "the gods of old" were really technologically advanced aliens who genetically engineered "Adam," meaning the early *Homo sapiens*, or Cro-Magnon man, why is there no physical, *high-tech* evidence of their existence? Why are no sophisticated alloys, machines, or electronics ever found in the archeological digs and ruins? And why were their "chronicles" written on clay tablets instead of in books or on ancient CD's? In short, why is there no tangible evidence of high technology—and thus of their existence?

Still, there is some compelling or at least tantalizing evidence. There is a rich legacy of the Sumarian and Near Eastern writings; there is the ancient architecture and astronomy which often evinces a high level of technical knowledge. There is evidence of gold mining in southern Africa, pre-dating the earliest known civilizations. There are inscriptions, paintings, and carvings which really do look like astronauts, rocket ships, and solar systems. There are convincing stories and "chronicles" about the gods' creation of humans from more primitive species; and much more.

Nevertheless, the idea which is most intriguing posits the "Nefilim" or "Anunnaki," meaning the alien "gods", as the likely creators of civilization—including agriculture, husbandry, the sudden appearance of the "naked ape," (modern man), the sudden knowledge of many arts and sciences; and much more—all of which seemed to appear "whole cloth," out of nowhere, in the Near East a mere few thousand years ago. But I would be remiss if I did not also mention that there is good scientific evidence, though scant, that man evolved naturally (see Chapter 2), and that the "sudden" appearance of high civilization may have a better explanation (see next chapter). Nevertheless we will proceed according to plan and give Sitchin his hearing.

The word "Nefilim" means in Hebrew, "those who fell," but is usually translated in English Bibles as "giants." Sitchin believes that they were the "gods" described in the Biblical Book of Genesis, and that they actually existed in the ancient Near East, but were in reality aliens from another planet; and moreover that they later became the self same gods of Greek and Roman mythology.

One of the reasons Sitchin's theory is so intriguing is because it is as if he answers biologist Michael Behe's riddle, (who were the designers?). The 12th Planet also provides a curious connection to astronomer Tom Van Flandern's Exploded Planet hypothesis, and his search for "Planet X," (Van Flandern 1993). But Sitchin's "12th planet" did not explode, and has the unlikely elliptical orbit of a comet, with an orbital "year" of over three thousand years!

Overall, the book is questionable scientifically on many grounds but tantalizing on others; nevertheless it, along with books like *Chariots of the Gods?* (Von Daniken 1970), are great examples of out-of-the-box thinking. There are fundamental tenets of our present "dogmatic slumber" which it is not at all unreasonable to challenge or question. One is the idea that we are "alone" in the universe. It follows that if other intelligent beings exist, it is not unusual to think that they have visited us. This reasonable assumption, however, does not constitute scientific proof.

ON THE ORIGINS OF THINGS

We need to return briefly to the subject of Chapter 2, namely, Michael Behe's challenge to Darwin's theory of evolution by natural selection. I hope that we came away from that discussion with the idea that no serious scientific theory can ignore Darwinism. It is much too logical and it explains too many facts — but not all. It doesn't solve the problem of "irreducible complexity." It doesn't sufficiently explain how life began on Earth in the first place. It doesn't explain the sudden appearance of myriad species at different periods, or the gradual progression from simple to more complex life forms, or certain mass extinction events such as the one at the K/T boundary of 65 million years ago. It may not sufficiently explain the origin of *Homo sapiens*. For these answers, we may have to look elsewhere. We will also discuss some of these issues further in Chapter 14.

One of the things we have to look at is the possibility of intelligent life on worlds other than our own small Planet Earth. To do this, we can not yet rely solely on scientific or empirical evidence to support our claims. Although there is some, (see Chapters 11 and 14), it is still tentative. But we can rely on some compelling logic, and we can extrapolate from conclusions drawn by other Apocryphal scientists.

BEHE'S ELEPHANT

Behe used the idea of an "elephant" to describe an idea which is so big, and so obvious that sometimes very small creatures in the neighborhood, meaning us, fail to even notice its presence. That concept has some relevance to the present discussion.

That the Earth is a globe in space — rather than a flat disc with a "firmament," or "sky dome" over it, held up by Atlas — should have been obvious, but wasn't. That the wonder of life as something much more sophisticated and complex, especially at the molecular level, than some simple mixture of basic elements—should have been obvious, but wasn't. The impossibility of a human civilization springing up "whole cloth" with all of the accouterments and trappings of high civilization—where previously there were only simple hunter gatherers—should have been obvious, but wasn't. These are all examples of Behe's elephant. But there are those who will never see it no matter how much we point to it. They hide forever behind their *a priori* assumptions and their narrow, special interests. This chapter (indeed this whole book) is in the spirit of pointing out Behe's elephant.

EXTRATERRESTRIAL LIFE

As previously mentioned, many scientists today accept the possibility that there are other worlds in our galaxy where life exists. Even if the speed of light is the fastest that any spaceship will ever be able to travel, (it may not be), there is still plenty of opportunity, given enough time, for travel and communication between worlds which are thousands of light years apart. We are theoretically capable of such travel ourselves, if we had the will to do it. I don't find it at all ridiculous to suppose that as I write this, such journeys are being made by someone, somewhere.

Although he may not have proven his case, Sitchin's thesis is not at all ridiculous. And although such ideas always raise hackles and evoke ridicule among mainstream scientists, all that should be needed for their acceptance, and to accept Sitchin as a bona fide scientist, is sufficient proof. He is not far from providing it, as are others.

At this point in time, I know of a few scientific trends or possibilities which would lend credence to Sitchin's line of reasoning. The first; the Face, and other artifacts, at Cydonia on Mars (see chapter 11), is to my mind the best and most promising, because sooner or later it will either confirm or falsify the hypothesis of intelligent extraterrestrial life. Another is the work of geologist Robert M. Schoch, who proved by indisputable geological evidence that the Sphinx in Egypt must be approximately 9,000 years old. His theory, which will be the subject of the next chapter, in itself does not prove the existence of ancient astronauts. But it does prove that the current paradigm—which claims that Egyptian and Mesopotamian civilization began no earlier that 4000 BC—is sorely inadequate and needs to be revised.

Another approach was suggested by Van Flandern and others. If tangible, and verifiable evidence can be found of advanced technologies in ancient times; such as radioactive residue from nuclear explosions near the Dead Sea, or from "nuclear reactors" in the pyramids at Giza in Egypt, such evidence would be hard to ignore. A fourth line of reasoning is the idea of "panspermia" or "astrobiology." This is the idea that life originated outside the confines of the Earth. This subject will be discussed in Chapter 14.

SOME TANTALIZING BUT INCONCLUSIVE EVIDENCE

Sitchin himself provides some possible evidence, none of which proves his thesis, but it does encourage us to at least give him a fair hearing. A sample of this evidence will be given here:

1. Ancient depictions of the Solar System which are fairly unambiguous, including the furthest planets and the moons of other planets, which Mesopotamian astronomers could not have known about some 3000 years before the invention of the telescope (Sitchin 1978, 205).
2. Depictions of "gods," and "eagle" men and women, with helmets, goggles, back packs, and other equipment (132,166). These depictions could of course be mythology, but if so what are the referents? Where are the real life examples? Primitive men might be able to imagine men with wings, or "angels;" but men with goggles, helmets, and technical looking equipment?
3. Knowledge of mathematics and astronomy in the ancient Near East which surpasses our expectations for a primitive, early civilization.
4. This includes certain knowledge of the Earth as a globe. This knowledge was later lost until it was finally revived in modern times (192).
5. A sophisticated calendar was in use from the earliest times in Sumer, circa, 4000 BC (186).
6. That required, among other things knowledge of precession of the equinox, which is the theoretical arc inscribed in the heavens by the Earth's poles in a period of 25, 920 years. The Zodiacal "ages" are based on this advanced knowledge. But men had that knowledge, given to them, says Sitchin, by the Nefilim, from the earliest of times (194–97).

CONCLUSIONS AND OPINIONS

We could speculate as to how much of Sitchin's theory we will accept as written. But that is not necessary; we will proceed as usual and let the reader decide. What interests me as an Apocryphal science enthusiast is that the gist of it may possibly be right. Time will tell just how much.

I am inclined to believe enough of Sitchin's theory, not in all the details, but in essence, to include him as one of my Principals. But the idea of a "12th planet" with a 3000 year comet orbit "year," and consequently, the belief that for the "gods," one "year" is equal to three thousand years for us, is scientifically ridiculous. My hunch is that the Nefilim, if they existed at all on Earth, existed at a much earlier time in prerecorded history, and that the stories and legends in the ancient Near East were already by that time ancient stories. Perhaps they came here millions of years ago from an exploded planet where the Main Asteroid Belt is now. Perhaps Mars was their moon and they had a base there, and that's why the Face is there. Also, as Van Flandern has speculated, and I agree, the Face faces skyward, instead of horizontally like the Sphinx in Egypt, because it was intended to be seen from space — possibly by us. We shall see.

Chapter Thirteen

The Sphinx: Mild-Mannered Geologist Makes Waves

THE RIDDLE OF THE SPHINX

Robert Schoch first came to my attention as the geologist who claimed that the Great Sphinx in Egypt was much older than is commonly believed. Some readers may be familiar with the story as it was presented in the TV documentary narrated by Charlton Heston. Schoch had been asked by maverick Egyptologist, John Anthony West, to examine the Sphinx. Being a sober scientist and not given to wild speculation, Schoch did not expect to find anything new. But what he found amazed him. As a result, it changed his life enough to qualify him as a "radical" and hence, as one of the Principals of my story. The detailed story of how Schoch arrived at the real age of the Sphinx, along with the science which supports his theory, is well documented in his book, *Voices of the Rocks*, (Schoch 1999).

Using standard non-invasive techniques to study the weathering patterns of the limestone rock which makes up the body portion of the Sphinx, Schoch was able to determine that they were caused by heavy and consistent rainfall. But how was that possible since North Africa has had a dry, desert climate since at least 3000 BC? In other words, the Sphinx has existed in a dry climate since its supposed construction by the Pharaoh Khafre around 2500 BC; so how could it have weathering patterns from heavy and persistent rains? Schoch knew that between 3000 and around 10,000 BC, when the last ice age was ending, the climate in that region was much wetter. Therefore, he reasoned, the Sphinx must have been constructed during that earlier time period.

The depth of the vertical crevices or "runnels" caused by many years of rain weathering, plus seismic data taken by Schoch's colleague, Thomas Dobecki, pushed the date back much further than the previously accepted date of con-

struction. Taken together, their calculations bring the date of the original construction of the Sphinx body to 5000—7000 BC or earlier, or between 7000 and
9000 years ago (Schoch 1999, 42). This places the probable date of the Sphinx's
construction at between 2500 and 5500 years *earlier* than the conventionally accepted date! That means that at the time of the ancient Pharaohs of Egypt, the
Sphinx was already an ancient monument. But who then built it?

Before I go into any more detail, we should ask why it is extraordinary or
"Apocryphal," to pose such a hypothesis. The obvious reason is that it challenges a major, entrenched mainstream theory—that of Egyptologists and anthropologists; and this always presents problems in today's world of power
politics in science, as we have been seeing throughout this book.

According to the existing model or paradigm, prior to 3500 BC humans
were much too primitive to build the Sphinx. Although they had already invented agriculture and had begun to domesticate animals, they were still
"stone age" people, and as such had not yet developed the skills and characteristics of high civilization, the kind needed to make large construction projects like the Sphinx possible—that is, not until the founding of the Sumerian
civilization, and soon after that, the Egyptian. If there were high civilizations
before that time, say Egyptologists, where is the evidence? This is a question
we will deal with in this chapter.

But there is much more to it than that. By placing the construction of the
Sphinx perhaps several thousand years prior to the previously accepted date
of circa 2500 BC, we open up a whole new set of questions. We open up whole
new vistas and new possible "world views." We make a fundamental change
in our assessment of humanity and its origins. All are sufficient reasons for
establishment science to reject Schoch's findings. This issue will also be discussed in due course.

Schoch offers some sober scientific evidence in this regard, but he is no radical, nor is he a "New Age" theorist of the type discussed in the previous chapter.
But what he proposes may be equally outlandish, and perhaps even more threat-
ening to mainstream science than any theory of ancient astronauts. The reason is
that, on scientific grounds, and by the rules of mainstream science itself, his argument carries a lot more weight than New Age theorists like Sitchin, and can't
easily be brushed aside. His critics fear that he may very well be right. He may
well succeed in helping bring about the "paradigm shift" he seeks.

ON PARADIGM SHIFTS

Thomas Kuhn's idea of paradigms in science and how they come to change
over time is now fairly well known (Kuhn 1970), and it has already been

mentioned in these pages. Like Ockham's razor, it has entered into the common vocabulary or parlance of science—so much so that it has been cited by several of the Principals of this story in their writings, including Schoch. Since the idea has become almost common knowledge by now, I will stress again that it parallels a theme of the present book. The idea is that once a scientific theory becomes a paradigm, or in other words, an established or mainstream belief, it becomes almost an article of faith, and is therefore very difficult to challenge or change. It can be done, but it is difficult—for all the reasons we have discussed already in previous chapters.

But there is a difference between Kuhn's idea and mine, not mine exactly since I learned of it from the writings of Ayn Rand although in a different context. Kuhn considers the idea of paradigms, and the difficulty of changing them, to be in the nature of the scientific enterprise and for the most part specific to it. That is true—as far as it goes. What I have added, is to show that there is another dynamic operating also, something with a more general or universal application. It is something that can precipitate declines in civilizations, and sometimes brings an end to historical ages of scientific progress.

The idea is that with the growth of government and its bureaucracies, (as with the established religious organizations of the past), comes government support and funding of "acceptable" theories of science. In this environment, the various branches of mainstream or establishment science create their own canons and dogmas, and acquire greatly enhanced power to eliminate competing ideas and their advocates—by force if necessary. A more subtle but no less important factor is the philosophical or ideological "paradigm" and social milieu under which the scientific model is subsumed. This larger philosophical context favors theories which it finds compatible with its larger "theme." All this makes challenging or changing scientific paradigms much more difficult than it would be in Kuhn's model, and explains why free societies progress while dictatorships and theocracies do not, relatively speaking.

In many ways, we are at that stage now, with one exception. Since free speech still exists, dissenting theories are still possible, although they succeed only with great difficulty—and hence can be considered Apocryphal.

Schoch does not go so far in his comments on the subject. Still he sees, by Kuhn's criteria, the difficulty of changing the current paradigm. But as I already said, I think he has a good chance to do it. For one thing, he's got the "goods," meaning he has a viable theory, good evidence, and lots of mainstream allies. For another, I don't think the Egyptologists alone have the power to stop him, but I could be wrong.

ABOUT ANCIENT ASTRONAUTS

If Schoch is right in his estimation of the age of the Sphinx, and I think that there can be little doubt that he is, it means that either human civilization is very much older than is now believed, or human beings had help from alien "gods." What is *not* possible is an idea that I have expressed previously. In my opinion high civilization could *not* have sprung up "whole cloth," out of nowhere, circa 3500 BC, as stated in the last chapter. There would have been far too much for humans to learn from scratch, and they could not have learned it so quickly. But this is just what the mainstream theory implies.

Schoch believes that civilization is old enough to have had plenty of time to have acquired all of the knowledge and skills exhibited in ancient Mesopotamia and Egypt, and to have been able to build the Sphinx eight or nine thousand years ago or more. But I think the jury is still out on the issue of "the gods." Civilization may indeed be very old but that does not rule out the possibility of alien influence, for the reasons given in past chapters, and in one to come. Still, if I had to make a choice right now, I would side with Schoch. You can't argue with the facts, and Schoch has them—or if not that, at the very least he has a lot of good evidence.

CIRCUMSTANTIAL EVIDENCE

There is of course circumstantial or hearsay evidence on both sides—that supporting the conventional age of the Sphinx, and that opposed to it. But we are concerned here only with the opposing or Apocryphal theory.

In ancient times, it was widely believed that there was some mysterious "riddle" connected with the Sphinx ((Schoch 1999, 33). It seemed to have had to do with its great antiquity, even to the ancients themselves. In the last century, the translator of the *Egyptian Book of the Dead*, E. A. Budge, believed that the Sphinx was older than Khafre. Sir Flinders Petrie, a famous Egyptologist and a founder of that science, also thought so (43). More recently a credible New York City police forensic expert, Frank Domingo determined that the head of the Sphinx did not match a known bust of the Pharaoh Khafre; and he concluded that they were not renditions of the same person. In any event the present face and head on the Sphinx is probably not the original, as it seems to have been re-carved several times (44–5).

Some other circumstantial evidence pointing toward an older age for the Sphinx is as follows: An inscription dating from the sixth or seventh century BC, called the "Stela of Cheops' Daughter," indicates an older origin. Also ancient Roman, Greek, and Egyptian references to the Sphinx, and the present day oral traditions of the local residents, likewise all attest to its great antiquity (Schoch 1999, 45). Since there is at least an equal amount of this type of evidence on the other side, it is fair to say that judging from circumstantial evidence alone one could not really say which theory is the correct one. What is telling in this case is that after the controversy stemming from Schoch's research erupted, and better evidence was presented, many geologists were eager to accept his findings, while Egyptologists continued to put up stiff, if not violent resistance (46).

GOOD EVIDENCE OF LOST CIVILIZATIONS
AND THEIR KNOWLEDGE

Regarding the demand of Egyptologists for evidence of earlier civilizations, there *is* in fact good evidence, but probably not as much as we would like to see. For archeologists, there are certain difficulties involved in getting at the evidence of lost civilizations. Some of it may be buried under yards of river bottom mud at unknown locations; and some of it may be under the sea, as we will soon see. The ending of the ice age around 13,000 — 10,000 years ago launched a period of heavy rains and flooding from the melting of the glaciers which raised the sea level some 200 feet. This would have flooded and destroyed any seacoast and river delta communities or cities which might have existed prior to that time (Schoch 1999, 55). Because of the difficulties that searching in these locations poses, archeologists have had to look elsewhere.

In Upper Egypt, for example, a chambered mineshaft for stone mining was radiocarbon dated at 31,000 BC (Schoch 1999, 56). That is considerably earlier than the standard model allows for such skilled activities. Nearby, sites dating to 8000 BC were found where sophisticated, well-planned villages, complete with wells and houses, were located. One village, called Nabta, was found with paintings and artifacts indicating cattle worship, and more interestingly, with huge quarried, nine foot high stones or "megaliths" standing on end. These were of the type found at Stonehenge, England, but dating a thousand years earlier. The fascinating thing about these megaliths is that they demonstrate a relatively advanced knowledge of astronomy, with slabs aligned exactly north-south, and east-west, and others aligned toward the rising Sun of the summer solstice, and casting no shadow at noon on those days (57–58).

Then there is the city of Jericho, located in ancient Israel / Palestine, dating to 8300 BC, with its huge stone wall, and equally large moat, excavated out of solid bedrock. And in what is now Turkey, dating to 7000 BC, was a community of several thousand inhabitants, called Catal Huyuk. This was a rich, orderly city, with paintings, sculptures and religious symbolism (Schoch 1999, 59). All of these examples and many others go well beyond the level of sophistication usually thought of as Neolithic, in the present paradigm. As Schoch puts it: "These sites were the work of people who knew what they were doing because they had been doing it for a long while, far longer than the old model of the rise of civilization allows" (59).

Were these people capable of creating the Sphinx? Let's look further.

It's quite one thing to look at the famous cave paintings, at Lascaux in France, dated to 15,000 BC, and say that they represent a primitive religious symbology. The theory is that the paintings were a kind of incantation or prayer that the hunt would be successful. That in itself would be impressive, but the truth may be much more so. One well-known mural is a case in point. The painting is known as the "Hall of Bulls." It is an array of various animals; bulls, and some other imaginary animals—types which were not hunted by these people, whose game was mainly reindeer (Schoch 1999, 62). So why did they paint pictures of animals they didn't even hunt?

Schoch cites a radical theory offered by Frank Edge, a high school teacher, which makes much more sense. According to Edge, the picture was of the summer sky of 17,000 years ago. The bull represented *Taurus*, "The Bull," one of the signs of the Zodiac! (Schoch 1999, 62). This theory, if true, totally unhinges the mainstream dogma that Paleolithic humans were simple hunter-gatherers who wore skins, went barefoot, and could barely speak, let alone understand astronomical concepts. They apparently had already invented the Zodiac, or were given this knowledge by the "gods," at least 17,000 years ago. "So striking is the resemblance of this ice age bull to the traditional picture of Taurus," Edge writes, "that if the Lascaux bull had been discovered in a medieval manuscript rather than on a cave ceiling, the image would have been immediately recognized as Taurus," (63).

Amazing as this is in itself, there is much more involved here than simply observing the sky and creating images to help remember the positions of the stars. But we can not go into any more detail in this brief summary, except to say that Schoch is convinced from this and other evidence that humans were much more sophisticated and knowledgeable in the ways of science, at a much earlier time in history, than is currently believed. One more example will suffice to make the point.

We mentioned precession in the previous chapter, and implied that from Sitchin's point of view it would have been highly unlikely for primitive humans to have discovered a phenomenon which takes extremely long periods of time to manifest itself. It takes over two thousand years for one "North Star" to be replaced by another. It takes a similar amount of time for one Zodiacal constellation to replace another as the leading constellation of the spring equinox, and thus to usher in a new "age." Since this knowledge evidently existed in ancient Sumer, Sitchin believed, with some good reason, that the knowledge was a gift from the "gods."

But what if humans had tens of thousands of years to develop this knowledge? What if the cave paintings were merely a stage in a development which began possibly much earlier? Would it then be feasible? Schoch thinks this is just the case. Edge's work, and similar evidence provided by others, lends credence to this opinion. Another piece of evidence is the "double ax," a symbol commonly associated with precession, which was found in shrines in Catal Huyuk, mentioned earlier, as were depictions of the planets and the twelve signs of the Zodiac (Schoch 1999, 74).

ATLANTIS

There have been innumerable books, many were written in the nineteenth century, about Plato's fabled lost city of Atlantis, which Plato claims to have existed around 9500 BC. Locations for this famous sunken city, or "island nation," have been postulated to have been at many various locations—ranging from islands near Greece, in the Mediterranean—to islands, or even a sunken continent, in the Atlantic Ocean—to America itself. Schoch gives these theories a new infusion of scientific plausibility, and a more reasonable interpretation. He reviews and evaluates those theories which seem to have at least some scientific evidence to support them.

One is that Atlantis was located in what is now Antarctica, the frozen continent at the South Pole. The story stems from a map dated to 1513 and attributed to a Turkish captain named Piri Reis (Schoch 1999, 95). It was revived more recently by science historian, Charles Hapgood (96). Hapgood was interested in Reis's map not only because of the anecdotal evidence about the map's sources, but because he came to believe in the theory that the Earth's crust can shift in a somewhat similar manner to the accepted theory of continental drift (97). If that did indeed happen, then it would be possible for this "Atlantis/Antarctica" to have once been located in a warmer climate.

Hapgood supports his theory not only by the Reis map, but also by several other maps subsequently found, some of which are more credible than others; but all, in my opinion, amount to nothing more than circumstantial evidence. A scientist can no more build a scientific theory based on them, than he can build one based on Greek or Sumerian mythology, and Bible stories—as Sitchin does.

Schoch needs more—and he finds it. One group of geologists, for example, has found evidence that Antarctica once had a warmer climate (Schoch

1999, 101). Schoch himself adds considerable weight to this evidence by giving serious consideration to the theory that comet or asteroid impacts with the Earth could cause "crust slippage, or "polar shift," (155–58). But all of that is a long way from proving that Atlantis was in Antarctica.

Another candidate for the title of lost city of Atlantis, is Yonaguni, another reputed, "lost city," this one located off the coast of Japan—under water. Schoch was again lured into examining the facts of the case by John Anthony West. And this time he had to learn to scuba dive to do it. In another televised special, Schoch examined this shelf-like underwater structure. His conclusion, disappointing to everyone, but proving he is a true scientist, was that the structure was probably natural, and not man made (1999, 113).

Schoch is open-minded enough to also consider the extraterrestrial possibility. But in the final analysis, he finds this line of reasoning to be "pseudoscientific" (1999, 114). And so would I, if it were not for the fact that there are a number of things which keep the possibility alive, which make the "extraterrestrial connection" yet a testable hypothesis, as we have seen and shall see in the next chapter.

EARLY EUROPEAN CULTURES

One archeologist who provides an important piece to the puzzle is Mary Settegast. In her book, *Plato Prehistorian*, she makes the case that the "Magdalenian" culture, which existed in Europe around 9600 BC, was the actual Atlantian race of Plato fame. This is the same culture which gave rise to the paintings of Lascaux and other cave sites in Western Europe. The people of this Paleolithic culture had a highly developed sense of artistic expression; they also tamed horses, probably the first Europeans to do so (Schoch 1999, 123). There is good archeological evidence that they took part in a 500 year long war, perhaps with the proto-Greeks—again, in keeping with Plato's story. Another fact that fits Plato's description is geological and archeological evidence that in about 7500 BC, very heavy rainfalls coupled with melting glaciers (126) had catastrophic results for this culture—an event which may have been handed down through the millennia as an oral tradition of the Sunken land of Atlantis, and other "flood stories."

What emerges is that these peoples were no Stone Age brutes; they were highly sophisticated, organized over a wide geographic area, had a knowledge of astronomy, and made sustained and coordinated war against their enemies. Could these be the people who built the Sphinx? Maybe.

Another view of an early European culture, that of archeologist Marija Gimbutas, is discussed by Schoch in his sequel to *Voices of the Rocks*. This work, called *Voyages of the Pyramid Builders* (Schoch 2003), is for the most part tangential to the present discussion, as it deals primarily with the question of how pyramids came to be so widespread over the world. A few of the issues he discusses in this book however, do have bearing on the present subject, and we will look at them where appropriate.

Gimbutas used the term "Old Europe" to describe a peace loving, goddess worshiping culture that lived in southeastern Europe between 7000 and 3500 BC. This culture gave rise to the Minoan civilization and after that, the Greek. Gimbutas believes that this culture invented writing, and was subsequently subjugated by a god-centered culture from what is now the Russian Steppes (Schoch 2003, 151). This is similar to an earlier theory by the historian, Alexander Rustow (Rustow 1980).

The point of all this is to show that there are several known possibilities for high culture in earlier civilizations, prior to what the mainstream paradigm considers the beginning, or "cradle" of civilization in Mesopotamia, circa 3500 BC. Could any of these earlier civilizations have been capable of building the Sphinx? I think what Schoch is unambiguously telling us, by his method of building a case based on a "preponderance of evidence," is that any one of them might have. But he never gives us any conclusive statement as to which one did—a fact which serves only to prove him an honest scientist.

A GEOLOGICAL PREPONDERANCE OF EVIDENCE

Throughout much of the two works mentioned, Schoch relies on his "home turf," (geology), and also sometimes astronomy, to build a good case through the method of preponderance of evidence. If no individual case can prove conclusively that early high civilizations existed, since an isolated case can always be labeled an "anomaly," there is no doubt, and much of the evidence in both books convinces us, that early humans had much more knowledge, and were much more sophisticated and accomplished than the present paradigm allows. If you put all of the scientific evidence together, the conclusion is inescapable that early high civilizations did indeed exist, and one of them surely built the Sphinx.

In addition to what we have already reviewed, an analysis of the geological and astronomical evidence is necessary for a detailed understanding of the how Schoch makes his case. This evidence will be mentioned only very briefly here. Schoch makes the case for a combination of glaciations and melting glaciers which caused, as was stated above, high rising sea levels. In addition, over thousands of years, intermittent comet and asteroid impacts with the Earth—some of which caused city-destroying tidal waves, mega-flash floods, and catastrophic rains—had a profound effect on the history and pre-history of civilization. These impacts affected the long-term climate in many various, and sometimes catastrophic ways. Radically changing climates may have radically changed the course of civilizations; possibly many times. This geological history was also the source, no doubt, of many a famous myth.

AN ANTEDILUVIAN CIVILIZATION—SUNDALAND

The most definitive statement Schoch makes regarding the true location of "Atlantis," or the antediluvian "Eden" of Biblical fame, involves such a climate change. It concerns an ancient, sunken sub-continent called "Sundaland." This information is found in Schoch's sequel book, *Voyages of the Pyramid Builders* (Schoch 2003).

Sundaland was located in Southeast Asia. It was a land bridge extending from Indochina, and almost reaching Australia. Except for the island chain presently existing in that area, including Java, Borneo, and the Philippines, it is now underwater (Schoch 2003, 245). The theory originated with a scientist named Arysio Nunes dos Santos, (244), and was updated and fleshed out by a physician named Steven Oppenheimer, in his book, *Eden in the East*, (246). Again, direct proof is scarce; it's hard to find factual evidence of lost civilizations under the sea. But the circumstantial, and even logical, evidence is compelling.

It is known that Sundaland was flooded three times in the "recent" geological past. The "high water" marks were approximately 13,000, 11,000, and 6500 BC, with sudden "cold snaps" in between where the ice age resumed (Schoch 2003, 242–43). Schoch thinks there is good evidence that the gradual processes which cause ice ages were speeded up by comet and asteroid impacts. Depending on whether they landed in the sea or on land, these massive impacts could have caused sudden warming, with torrential rains, tidal waves, and flooding—or sudden cooling (261–66), with the return of "mini-ice ages."

The theory is that this history of actual catastrophic flooding of the fertile lowlands of this ancient high civilization was the source of the "flood myth," and many other ancient myths, handed down through the millennia by numerous civilizations. The logic of the available evidence is simple. The myths—and certain other evidence—are like "spokes on a wheel." They are more concentrated, and / or more ancient, near the center, and more diffused, and thus recent, as they move away from the center. The center of the "wheel" is, by deduction, Sundaland. Following are some examples:

1. The best candidate for the original flood story, the one many of us know as the story of "Noah's Ark" in the Bible, seems to be India. This story of the original "flood hero" named "Manu" was most likely carried to Mesopotamia and Greece along trade routes (Schoch 2003, 250).
2. The language group of Southeast Asia is Austronesian. This language is, along with the Indo-European language group, one of the most prolific and oldest in the world, and it is centered in Sundaland.
3. The Austronesian speaking peoples have more flood stories than any other culture group (251). The same is true of other Bible stories and culture myths such as "Cain and Abel," the terrible serpent "Leviathan," and so on.
4. The *Egyptian Book of the Dead* mentions "Manu" in its legends of "olden times."
5. The location of Sundaland in the Far East coincides with the legends of many other peoples who put the location of "Eden" in the "East."

Although the factual archeological evidence supporting this theory is scant, it does exist, and future work may buttress it or refute it.

CONCLUSION

Schoch theory, as a whole, is a testable hypothesis, and hence, good science. For readers interested in theories of ancient civilizations, Atlantis, and Pyramids, Schoch is a good place to start. Schoch concludes, by the way, that the Great Pyramids are *not* older than they are believed to be in the mainstream theory, but that the *site* probably is—however, that's another story. Unlike many other authors in this genre, Schoch is never pseudo-scientific. He will get you off to a good start in your quest to challenge mainstream science on the subjects of the Pyramids and the Sphinx.

Chapter Fourteen

Space Seeds:
DNA: The Cosmic Genetic Code

TWO EXTRAORDINARY HYPOTHESES

Darwinism is often worn like a badge of honor by atheists, secularists and other intellectuals. Unwavering belief in the Theory of Evolution since the late 19th century has been the hallmark of the "rational" and "modern" man or woman. Judging from the reactions of some of those just mentioned whenever the subject of Michael Behe is brought up, my impression is that for them to give even an inch of credibility to Behe's theory is to go over to the enemy, to become a "true believer" or worse, a "conservative." Therefore Behe, and I suspect also the Principal of the present chapter, Rhawn Joseph, are *persona non grata* in that world. They can be given no quarter, no shelf space, and no acknowledgement that there may be matters of science to discuss here.

In this chapter, Joseph, a Ph.D. in neurophysiology, with his book, *Astrobiology, the Origin of Life and the Death of Darwinism* (Joseph 2001), challenges this formidable paradigm. He proposes two extraordinary hypotheses. He makes two cases; both are testable, scientific theories, and both are backed-up by good evidence.

Case # 1—There is an ample fossil record to say with some assurance that microbiological life forms based on DNA, the foundation of all life on Earth, have come to Earth from other worlds on meteorites, asteroids, and comets. There is also evidence that microbiological life once existed on the Moon, and that it once existed and probably still exists on Mars. This is a paradigm-busting hypothesis because if true, it falsifies the Darwinian theory that all life spawned from a single cell which spontaneously arose from the basic elements and molecules of the young Earth. It is a testable and falsifiable hypothesis because we now have the technology to go to Mars and even to the moons of Jupiter and Saturn, and we can accurately test for signs of life on meteorites which fall to Earth. There is good reason to suspect, and we will soon know without a doubt, whether or not life originated and exists outside of, and independent of, the Earth.

Case # 2—Joseph's second hypothesis concerns Evolution, another "sacred canon" in today's science. Let's be clear on one thing; although he advocates a kind of "design," Joseph is most certainly hypothesizing a scientific theory. There was, and is Evolution. It is supported by an ever growing fossil record. But Darwinism and Evolution are not synonymous. *Darwinism says that evolution is caused by the "natural selection" of random variations and mutations of existing species.* But Darwin's process may be only part of the story; it may do no more than select from predetermined changes which have already arisen according to the specific instructions of a very ancient genetic code—a cosmic genetic code no less.

Joseph says that evolution is caused by a pre-programmed "metamorphosis," and that the program is in the DNA. Just as a caterpillar changes into a butterfly in one season, species evolve or "morph" into more advanced species, over millions and even billions of years, when conditions are right. The process is intricately tied to the "genetic engineering"—again, the DNA does the engineering—of "Mother Earth," to make it suitable for more advanced life forms.

This is also a respectable scientific hypothesis, and it is eminently testable. This is an important book, and cutting-edge science at its best. Therefore, I confidently enter Rhawn Joseph into my pantheon of Apocryphal scientists. Here is another innovative theory which deserves a fair hearing.

GETTING FROM THERE TO HERE

As astronomers Halton Arp, Tom Van Flandern, and others have suggested, the universe may be exceedingly old—perhaps infinitely old. In any event it is certainly far older than the paltry 12–15 billion years or so, predicted by the Big Bang (BB) theory. Joseph subscribes to this "infinite age of the universe" theory as a matter of course, perhaps for philosophical reasons, and perhaps without knowing of the good evidence, such as Arp's, which clearly demonstrates the falsity of BB hypothesis.

In fact, an "infinite age" for the universe is a prerequisite for the rest of Joseph's theory. He postulates moreover, that *life*, and specifically the DNA which is the basis of all life, must also be very old, and must have existed on innumerable other worlds in other solar systems, and in other galaxies, long before the Earth was ever formed. There are good reasons for these assumptions, and they will be discussed below.

The mainstream or establishment theory for the origin of life on Earth, credited to Darwin, is that it arose from the "steamy cauldrons" of the newly formed Earth around four billion years ago (Joseph 2001, 25; Fortey 1997; Dawkins 1996). Joseph begins with the assumption that this would have been impossible, because DNA is composed of ingredients which did not exist on the newly formed Earth—ingredients such as free oxygen and phosphorous. In the newly formed Earth, these elements were tightly bound in minerals and not available for the formation of DNA (Joseph 26, 69, 132). Also, it has become increasingly clear to researchers that it may be impossible to synthesize DNA, even from the complex organic compounds which probably existed on the nascent Earth.

Alternative theories have been proposed by others. One is that at first, simpler organic proteins were formed in the "steamy cauldrons"; after that RNA, which is somewhat less complex than DNA, was formed from these proteins; finally DNA arose. Another theory is that complex organic matter, but not DNA, may have arrived here from outer space—formed perhaps in cosmic nebulae. But these theories are contrary to the evidence, says Joseph, and contrary to the newly expanding knowledge of the genome. "Only DNA can beget DNA, only life begets life." Joseph demonstrates this fact over and over throughout his book.

So where did the first DNA on Earth come from? Joseph, citing recent findings from the newly emerging science of astrobiology, or exobiology, postulates that it came, and is still coming, from outer space, "hitchhiking" here, as it were, on space debris which may be the remnants of long extinct worlds (Joseph 2001, 72). As is commonly accepted in astronomy and in mainstream cosmological theory, Joseph points out that during the first seven hundred million years after its formation, the Earth was constantly bombarded by space debris, comets, meteorites and asteroids (27). In other more radical theories, such as Van Flandern's Exploded Planet Hypothesis, the Earth was also intermittently bombarded anew when a planet, or planets, exploded in our solar system (see chapter 10). Even during relatively stable periods like the present, tons of organic, carbonaceous material is rained down on the Earth yearly from outer space (39). All this abundant matter, says Joseph, must have contained, and still contains, sufficient living matter in the form of bacteria-like, life forms, and other microorganisms. The genetic information contained in the DNA of these "cosmic hitchhikers" is the basis and origin of all life on Earth.

How does all this space-faring life survive the perilous journey in the frigid cold of space for interminable lengths of time, until such "space seeds" can reach safe-haven on Earth or any other world? Joseph goes to great length to explain how—which we can only briefly recapitulate here.

VERSATILE, TENACIOUS LIFE

As any cook knows, it is not always so easy to simply boil away bacteria in food. Nor does freezing or canning always protect food from microbial contamination. Sometimes more drastic measures are required. The reason is that life can sometimes be very hard to kill. It can develop into forms which are resistant to extreme heat, cold, pressure, and shock.

As an example, some bacteria evolved methods of encasing themselves in stone for protection. Joseph reports that most of the creatures or microorganisms which inhabited the primeval oceans of the young Earth, three and a half billion years ago, likely arrived here "encased in all manner of stellar debris, and were well adapted for surviving space travel and exposure to cosmic radiation" (2001, 70). "In fact . . . gel-secreting, photosynthesizing, cyanobacteria communities were constructing stromatolites or "stone mattresses", upon the surfaces of the shallow water seas." Fossils of almost identical microbacteria were found in Martian rock examined by NASA scientists. Moreover, geologists at the University of Glasgow report that over 200 square kilometers of white rock on Mars appears to be stromatolite (p. 71).

Given that the same type of life forms found on the nascent Earth seem to have been also found on Mars, Joseph surmises that the ancestors of such creatures "hitchhiked across chasms of uncharted space, clinging to cosmic dust and pieces of planet, and encased in . . . jagged blocks of ice" (2001, 71). These "hitchhikers" may have made their way through space for an unspecified length of time and through untold distances, and then pummeled the Earth, (as well as other planets), especially during its early formative period.

In 1864, a meteor landed in France near the town of Orgeuil. It was examined by the famous Pasteur who concluded that it contained fossilized evidence of microbial life. More recently, in 2001, it was examined again, and found to contain trace amounts of amino acids, the building blocks of life. Carbon isotope measurements indicate that these amino acids are of extraterrestrial origin (Joseph 2001, 73), and that they originated on a planet containing organic compounds and water (74). In 1969, a meteor scattered fragments across Murchison, Australia. These fragments were examined and analyzed several times using state of the art methods; by R. Brown of Monash University in Melbourne, by R.B. Hoover of NASA's Marshal Space Flight Center, and by Russian biologists, S. Zhmur, and L.M. Gerasimenko. All of these scientists determined that there was good evidence that the microfossils found in these fragments indicated that they were the result of biological activity—meaning that there had been living organisms within the meteor (75).

Joseph cites many other examples similar to the ones related above, but to review them all would be beyond the scope of the present book. The bottom line is that, as a whole—taking into account the possibility of contamination by Earth-bound organisms, considering the difficulty of perpetuating a hoax of this magnitude, given the distinguished academic credentials of the scientists who draw these conclusions—it is not difficult to conclude that life can, and has, originated outside of the confines of the Earth's biosphere.

Moreover, if as Van Flandern, Ovenden, Titius, and Olbers have hypothesized, (see chapter 10), planets do indeed explode from time to time, it is not at all difficult to imagine frozen oceans of ice and debris careening into space, teeming with organisms that are well equipped to survive this treacherous "star trek." Might not some of these organisms, Joseph asks, awaken once again on some distant hospitable planet ready and able to once again begin anew the evolutionary journey? (2001, 89). This "journey" consists in a predetermined DNA "master plan" to genetically engineer the new planet, and to evolve life according to a logical progression, through a process of metamorphosis.

Scientists have experimentally demonstrated that microorganisms can survive brutal environments of the kinds that such space journeys would entail. In one example bacillus spores were subjected to tremendous shock, equivalent to the pressure range which impacting meteors are subjected to—and they survived. Other tests showed that the interiors of some meteors do not exceed 40°C, which is well within the survival temperature of many bacteria. On the other end of the temperature spectrum, it is known that the interiors of space debris can be much warmer than the exposed exterior, for a variety of reasons such as pressure, insulation, chemical reactions, and not excluding biological activity (Joseph 2001, 89). In fact, many microbes can survive almost regardless of conditions. Species of microbes can survive boiling hot springs, frozen methane, antibiotics, tremendous pressure, shock, and even explosions. There are microbes which eat iron and thrive in salt rock (91), and there are organisms that can spring back to life after extremely long periods of dormancy. With this kind of evidence and logic, Joseph concludes that, living organisms, often including multicellular species, and possibly even insects (104), long ago arrived here on Earth from other worlds. They came, replete with the genetic information in their DNA, inherited from those previous worlds, necessary to begin anew a process lasting perhaps billions of years, of genetic engineering of the Earth and biological metamorphosis of species—culminating in "woman and man."

The remainder of this chapter will be devoted to a review of the fundamentals of Joseph's theory of evolution by metamorphosis.

ABOUT CELLS AND GENES

In order be able to form an opinion as to which is the correct theory—
Joseph's theory of evolution by metamorphosis, or Darwin's theory of evolution by random mutations and natural selection—we have to know something
about how the process works. The following summary may provide some of
the basics. The reader wishing to study the subject in more detail will have to
go to the primary sources and to current books on biology and genetics.

To begin with; two facts are presented which are supported by all known
evidence: 1) "only life begets life, and only DNA produces DNA." 2) "All
living cells are fundamentally alike chemically, structurally, metabolically,
and in regard to their cellular components" (Joseph 2001, 115). Further, all of
the processes necessary to sustain life, and to construct life in all of its many
various forms are contained within the cell and more specifically, in the
"macromolecules" or "genes" contained in every living cell—otherwise
known as deoxyribonucleic acid, or DNA. Even the simplest organisms, the
simplest single cell creatures, are incredibly complex. To Joseph, the idea that
they arose spontaneously, or even gradually on the young Earth is untenable
when we consider the unbelievable complexity of a single cell, and the fact
that even the earliest known cells were equally complex (116). Just how complex a cell is can be seen as we progress.

The simplest organisms contain around 6 thousands of these macromolecules or "genes" (Joseph 2001, 130); a fly may contain around 15 thousand
genes, and a human has 30 to 40 thousand genes. There are 46 chromosomes,
in 23 pairs, in humans. Again in humans, chromosome number 1 contains
several thousand genes, but chromosomes number 21, 22, and 23 are much
smaller, and the numbers in-between are medium sized (128). Bacteria contain only one chromosome (127).

There are, in all organisms, both *active genes* (exons), and *inactive genes*
(introns). Active genes comprise only a small fraction, about 3%, of the human genome. The rest, the vast majority, are *silent* or inactive. These are still
regarded as "junk genes" by some mainstream biologists, (see chapter 2). But
they are certainly not junk; they are genes which, when activated during the
process of evolutionary metamorphosis, will produce new traits, and new
species.

Each gene contains two strands made up of amino acids, which are themselves large, complex molecules or *nucleotides*. The two strands are held together with a "backbone" of sugar phosphate molecules containing among
other things, oxygen and phosphorus, (which, as mentioned above, were not
available on the young Earth). These strands of nucleotides are laddered together to form the now well-known "double helix" of the DNA molecule.

There are four kinds of nucleotides; for simplicity, we'll call them A, T, G, and C. The order or sequence of nucleotides determines the information or "genetic code," which determines the structure and function of all components, of all living things (Joseph 2001, 118). There are around *3 billion* matching or complimentary "base pairs" of these nucleotide sequences in the human genome (132). There may be as few as three, or as many as several thousand nucleotides in each sequence. So, as you can see, for humans there is an immensely large number, and for all species combined, there are an "infinite" number of possible sequence combinations. But probably there is a much smaller than an "infinite" number, but still a very large number of sequences that "work," meaning, that can produce viable life forms. This accounts for the variety of life.

When whole silent genes and silent nucleotide sequences or introns which may be part of already active genes, have the ability to make copies of themselves and to shift positions. These "jumping genes" or "transposons" often move to more active regions. When they change position they become active. This is not just Joseph's opinion. An assemblage of the world's most knowledgeable geneticist from the International Human Genome Consortium has drawn this conclusion and many others which will be related as we proceed (Joseph 2001, 118). These "silent genes" are ancestral genes passed down through untold generations and from previous worlds. Introns thus activated awaken long silent, but pre-coded, information which existed as Joseph says "*a priori.*" Thus they activate new traits and forms, and also create new species (118). This is not a "random" process as in Darwin's theory, but a "purposeful" process, meaning that it is pre-determined by chemical, biological and physical laws, and pre-coded in the DNA.

When single intronic nucleotides shift position; a "frame shift" results and the genetic code is altered. Further, "intronic genes may often contain genes-within-genes. Intronic genes may also make duplicates of themselves" (Joseph 2001, 119). As an intron becomes active it may become in the process a new gene. "Whole gene duplication" (WGD), so-called when intronic genes make copies of themselves, may be the means by which great bursts of evolutionary change occurs (119). "This process of creating new genes and duplicate genes is not random nor is it due to mutation, but is under genetic control and functions in accordance with precise genetic instructions" (120).

There is yet another possible cause of evolutionary metamorphosis; it is the process of inter-species transfer of genetic material made possible by means of *plasmids*. Plasmids are copies of nucleotide sequences which can exit a cell and "infect" another cell—or even another organism of the same or different species. Thus for example, bacteria may transfer newly evolved genetic material from one individual to another of the same species, to allow for the rapid and wide-spread metamorphosis of the new trait or species. This may also be responsible for "multi-regional metamorphosis," that is, evolutionary changes that occur almost simultaneously in several geographical locations. Although I find this method questionable—sort of akin to "Spiderman" being bitten by a spider and then acquiring the ability to spin webs and to climb tall buildings, it is an accepted theory in genetics—within limits (Joseph 2001, 120). Transfer of traits through plasmids allows species to evolve as one. If not more than one individual acquired some newly evolved trait, says Joseph, "it would immediately die out, alone and isolated" (122). Plasmids allow large numbers of individuals to acquire the new trait at the same time.

Parents and their offspring share the same identical genes. They do not, however share the same chromosomes. Chromosomes are constantly shifting from parents to each individual offspring. In humans for example, one child may be blond with blue eyes, another may have dark hair and eyes; another may be smart, another tall, etc. This is because the combinations of identical genes vie for dominant positions within the chromosomes. This is what allows for variation within species and even within families (Joseph 2001, 129). It does not however, allow for new, previously non-existing traits, or new species. To accomplish that, there must be changes in the genes themselves. According to Joseph, that can only happen through evolutionary metamorphosis.

Chromosomes are short lived, but genes, according to Joseph, may "live forever" if they are faithfully reproduced (2001, 129). Actually, of course, genes don't "live forever" either. In the first place, every new gene is a *reproduction* of an older gene, and although it may be an exact duplicate of the old gene, it is not the *same* gene. It is made of new components, although the two may be chemically and biologically identical. Secondly, over time, as each metamorphosis occurs in a gene, although it remains essentially the same as all other DNA, there are now differences in detail; there are changes. Lastly, metaphysically, this throws another "monkey wrench" into any idea of a materialistic "eternal life." Only atoms may seem to "live forever," but even that may not be true, (see chapter 9).

MOLECULAR CLOCKS

As we have seen, DNA contains the genetic instructions for creating all of the varieties of life. It also contains the instructions for genetically engineering the process. To Joseph, there are no random processes, no accidents of evolution, except in the case of variations in chromosomes as explained above, which result in variations within a species. Mutations do arise occasionally but they usually result in the death of the mutated individual. Joseph's system implies "design," or more accurately, "causality," just as other scientific theories do; and it explicitly accepts the idea of "progress." All of this, of course, is contrary to the mainstream theory of evolution according to Darwin. This is no slight difference from Darwin's system; this represents a major departure—a new paradigm. This also allows, in principle, Behe's theory of irreducible complexity to be valid (see Chapter 2) in that new irreducibly complex organs or functions can arise through one, or a series of pre-coded genetic changes.

"Genome novelty," meaning *evolution*, is caused by modification and duplication of older genetic material. But the creation of new genes by this means is under strict regulatory control. According to Joseph, there is an obvious progression from single celled organisms to humans, and evolution has come about in a "step-wise" fashion, as if programmed by a "genetic clock" (2001, 154). Darwinism denies the idea of "progress," and instead postulates random, purposeless, change which only seems to be progressing, "by accident" as it were.

The evidence Joseph provides for his theory of step-wise progression is technical, nevertheless an example can be given here in an abbreviated form. One important protein, cytochrome c, which participates in oxygen utilization, is made up of some 100 amino acids in humans. However, as we move down the "evolutionary tree" from humans to the lower organisms, this protein becomes gradually less complex in a step-wise fashion. In primates, for example, 99 of the amino acids are identical to those in humans; in other mammals, 89 are identical; 86 in reptiles; 79 in fish; 69 in silkworms; 57 in wheat; and 55 in yeast (Joseph 2001, 156). According to one geneticist, "the rate of change in many genes is regulated by a clock which seems to tick simultaneously in all branches of the tree of life" (156).

The expression of this molecular or genetic clock requires that the Earth's climate, atmosphere, and oceans be genetically engineered to prepare the Earth for those species which have not yet evolved.

GENETIC ENGINEERING OF EARTH

Darwinism and mainstream genetics accept of course, the idea that the Earth was bioengineered to allow later life to survive. Organisms capable of photosynthesis such as plants, for example, produce the free oxygen in our atmosphere which allows animals including humans, to live. But the evolution of this process is considered by Darwinism to be random and accidental. To Joseph, the Earth was "terra-formed" purposefully, by pre-programming in the genetic code, and to continue to accept Darwinism in view of current knowledge; of the genome, of the fossil record, and of microbiology; is to accept an inordinate number of "coincidences" as statistically possible. Joseph rejects the notion of "coincidence" as tantamount to "junk science" (2001, 161).

A brief look at the fossil record makes Joseph's point clear.

Three and a half billion years ago, anaerobic, carbon eating microbes and blue green algae began to prepare the planet. Many were oxygen secretors and able to resist harmful cosmic and ultraviolet rays by various means. Just as oxygen was secreted as a waste product by many life forms, in the next billion years, as the Earth cooled, it was "prepared" in other ways. Ediacaran fauna developed which secreted calcium carbonate, which made it possible for later creatures to form hard shells, and later endoskeletons—bones. And curiously, once their "mission" was accomplished, these Ediacarans disappeared from the scene (169).

Moreover, after a billion years of microbial growth on Earth, huge "microbe mats" covered the Earth which served as food for later, more advanced species. In a step-wise fashion, the Earth was being prepared to sustain more complex life forms which could not have existed before. And Joseph believes that there is a built in genetic feed-back loop which allows the tumblers of the genetic clock to make progressive advances creating ever higher, more advanced life forms whenever the Earth or the environment is ready for them—culminating in "woman and man" (2001, 170).

Since we use only 3% of the human genome, and the rest of our genetic information is in silent genes or introns, who knows where human evolution will end? But that's another story.

Chapter Fifteen

The End of Science?
The Dead-end of Mainstream Science

OPTIMISM VS. PESSIMISM

What if everything you thought you knew turned out to be wrong? Imagine if you could fast-forward 1,000 years, but unlike in H.G.Wells' story, *The Time Machine*, civilization did not destroy itself, but instead progressed steadily during that time. How many of the beliefs held to be "scientific fact" today would still be true in that future time? My guess is that, on the theoretical level at least, some would not. Some theories, like Newton's Laws, would still be valid as approximations. But others, like Einstein's Special and General Relativity, would have long ago been moved onto classical grounds, quaint relics of a distant past. So would the theory that AIDS is caused by a virus, so would the theory of human induced global warming, so would the science behind the myriad of "miracle drugs," especially the psychiatric ones, consumed by millions today; and many more.

Science based on new theories and paradigms may have given us by then a cure for cancer, unimaginably long life spans, faster than light travel, anti-gravity technology, wireless transmission of electrical energy—eliminating the need for "the grid" which now plagues us—and much more. Colonization of many other worlds may be a reality, and perhaps even good music will make a come-back. But of course this is all still a long way off, still the stuff of science fiction. But maybe someday it will be so.

We have reviewed some of the theories and their supporting evidence, which may lead to that kind of a future, offered by the Principals of our story. They are the Apocryphal scientists of today, but perhaps they will be the legends of tomorrow, these scientists who have been brave enough to have proposed alternatives to current wisdom—and who took the heat for doing so.

In his book, *The End of Science: Facing the Limits of Knowledge in the Twilight of the Scientific Age*, John Horgan (1996) considers conventional science as The Final Word, the End-All and the Be-All, in his fancied "twilight of science." As was mentioned in the Preface of this book, Horgan believes that any speculation to the contrary is "ironic science," which he likens to pointless literary criticism and science fiction. But it is not always that; there is good science behind the vision portrayed above, and behind almost every Apocryphal scientist's theory presented in this book.

Horgan, a science writer for *Scientific American*, is a good soldier for the mainstream scientific establishment. He now sees his science in a malaise, coming to an end, as it were. Well, in some areas at least, it probably is. Like the guardians of knowledge in the Medieval Church, Horgan guards the modern day dogmas of science jealously, turning a blind eye to any new and promising theory, even those which by all logic seem to be irrefutable. Yes, his science may be ending; but if it is, it's a self-fulfilling prophecy. That may be the real irony—and a real tragedy if the replacement theories fail to be heard.

JOHN HORGAN'S IDEA IN SOME PERSPECTIVE

Since I disagree profoundly with Horgan's premise, I obviously do not consider him to be one of the Principals of this book. I am including him mainly as a contrast, to show what can happen if the mainstream establishment succeeds in stifling or killing every new scientific idea which challenges its canon law. That certainly has not happened yet. At this point it is still only a fear (or a hope?) in the minds of men like Horgan. But his idea needs to be dealt with nevertheless, and the challenge met, perhaps to keep it from *becoming* a self-fulfilling prophecy.

Horgan interviewed many famous contemporary scientists for his book. I must point out to be fair, that most of them do not share his dismal vision. Since most of his subjects are mainstream scientists, along with a few philosophers, I don't agree with some of their basic assumptions. Still they are scholars, and as such they stand heads and shoulders above Horgan, who is essentially, with all due respect, a philosophical pipsqueak, a prophet of doom, and a false prophet at that. It might be interesting to review what these scientists have to say about Horgan's premise, but this book is not about mainstream science. If you want to know their ideas, you can read them as I have.

THE IDEA OF PROGRESS

As was discussed previously, the idea of scientific progress and the belief in science in general, is primarily the result of the Enlightenment of the 17th and 18th centuries. There had been periods of scientific progress before, in ancient Greece, and perhaps in some "Atlantean" period of ancient history or pre-history. But since the Enlightenment era, the idea and reality of Progress in science, has been for the most part constant for the past 300 or so years. As was also discussed previously, at the end of the 19th century there was a belief in some quarters that science was then coming to an end, but that belief was soon dispelled at the dawn of the 20th century. Now that doubt, that crisis in confidence, is again rearing its ugly head. Can it be true this time?

I say no, not a chance, not yet anyway.

First, let's ask a philosophical or rhetorical question—one that can never be answered with finality. Could there ever be an end to science? I mean, is it possible? The answer depends on whether there is a limit to the amount of things it is possible to learn, it also depends on the ultimate nature of the universe we live in. Since we may never know the answer to these two questions, neither may we ever know the answer to the first. To smugly say; "Oh, yeah, the universe began 15 billion years ago with a Big Bang. It began as a singularity" . . . and so on; is not much different from reciting some mainstream science version of the "Apostle's Creed." "I believe in Darwin, Freud, Einstein, and Schrödinger, and Neils Bohr," . . . Not that each did not make some legitimate contributions to science. It is believing *everything* they said, on faith as it were, and thus closing our minds to the possibilities, that presents the problem. We can go down that road if we want to, but we are gullible if we do.

Instead, I prefer to look at the question in a more practical way; historically you might say. We, that is mankind, have often come to the "end of science," at least for awhile. Sometimes the level of mankind's acquired sum of knowledge remains static for very long periods of time. Paleontologists, for instance, say the proto-man, our hominid predecessors, used crude stone implements which didn't change much for two million years. Thus it is fair to say that their knowledge didn't progress much during that period. During the Middle Ages, again, it's fair to say that knowledge didn't progress much for a thousand years. And there were other similar periods. That doesn't mean that these were times when there was nothing more to learn, or that we were incapable of learning anything more. It is just that, for one reason or another, progress is not constant, and neither is it guaranteed. Mankind increases its knowledge in spurts as it were, in fits and starts.

It's possible that we are coming to one of those leveling off periods, if not right now, then possibly in another hundred years or so. If so, that still says nothing about the Big Picture. It does not answer the big question or questions. Horgan talks about, and seems preoccupied with, a "Big Question" of his own: namely; *why is there something rather than nothing?* Why does the universe exist at all? But even asking that kind of question is presumptuous, philosophical, and rather ironic. We really have no way of knowing whether one or the other state (something or nothing) is the "natural" one or "logical" one. Better to simply accept existence as it is, and go from there. Learn as much as we can, for as long as we can; that's what knowledge is all about. That's what I'm for.

I'm optimistic about the future of science, and I guess that's what separates guys like me from guys like Horgan. Sure we could blow ourselves up, but if we don't, we'll keep learning.

I want to end this chapter with another "what if." What if we, in some distant future, learned all there is to know? For that matter, what if right now, as I write this, there are super beings or "gods," who know all, or nearly all, that there is to know. I say, so what! *We* still have to learn it. *Every child* ever born has to learn. The struggle and the quest continue.

Chapter Sixteen

Real Conspiracies:
The Case for Apocryphal Science

OUR DEBT TO MEN AND WOMEN OF CREATIVE GENIUS

In her novel, *The Fountainhead*, Ayn Rand's hero, Howard Roark uses the words, "this will be my testimony and my summation" (Rand 1952, 677), as a prelude to defending his actions in dynamiting the Cortlandt Housing Project for the poor which he himself had designed but which had subsequently been corrupted by others—by mainstream architects and government bureaucrats. Then in a few short pages, he justifies his actions, explains his personal philosophy of life, and instructs the jury, the court, and the reader on the meaning and purpose of the human enterprise, and on our undying debt to all men of creative genius (677–83).

Although he doesn't quite convince the reader, he does succeed in convincing the jury. He is acquitted, gets the girl, and the story has a happy ending. Well, life isn't exactly like that.

In our story, we told of several real life men and women of creative genius, who if they are correct, could also be great benefactors of mankind. But, as we have seen, there are guardians of the status quo, who, like the Churchmen of old, if they had their way, would prefer that no-one ever knew of the existence of these Apocryphal scientists and their theories. Some of the Principals have nevertheless succeeded on their own in having their cases heard. Others are losing. The degree of their success or failure as time goes by will depend on many factors.

I have done my best to make their case, but now, in the next few pages, I will make my own "summation," so to speak. In doing so, I hope to help the reader to better understand why it is so important that these scientists and others like them are heard. They *will* have a chance to be heard if you the reader

146

demand it. Your future—our future—may depend on it, perhaps not on these particular scientists, but surely on what they represent, which is their spirit of free inquiry and this plea for the freedom of dissent.

ANOTHER APOCRYPHAL THEORY

One of the realities of writing a book of this kind is the inevitable harsh criticism it will engender. Even some of the Principals themselves will no doubt find one or more of the theories presented here to be beyond the pale, and will not want to be included in the same book with some of the other "more extreme" Principals. There is always the possibility that this book will be ignored and then cosigned to the trash heap—an Apocryphal theory about Apocryphal science. I hope that doesn't happen, but when you challenge any prevailing idea, you can be sure that the keepers of that idea will do everything in their power to shut you up, and to shut you down. If they can't do that, they will attack you with a vengeance.

I expect that and accept it and I'm quite willing to let events take their natural course. That this book exists will be enough. Even if it fails it will succeed, *if*, and this is a big *if*, the theory is correct. In the meantime, as long as people have the opportunity to see what's out there, to see the possibilities, to hear things they have never heard before; that will be a beginning.

THE PRINCIPALS REVISITED

In concluding this book, I will give my own general opinion on each of the Principals' theories. The tool I will use here will be nothing more than my own sense of intuition, and plain common sense. Since I have already done my best to present the evidence and logic of the Principals themselves, and sometimes to give my own interpretation of their science, there is no reason to repeat that process here. I already made their case. What I want to do now to summarize is to render a simple kind of judgment of the various theories, by way of tying up loose ends.

Michael Behe: The consensus among many thinkers and intellectuals I've spoken to is that Behe really is making a case for divine creation, and merely disguising it as a scientific theory. There may be some truth to that contention, though Behe won't admit it. In his defense I'll say this: being a competent scientist in his field, he has become aware of a valid and pertinent question which deserves an answer. Does irreducible complexity really exist, and if so how does it affect the traditional Darwinian model? I think it is too early to say for certain, but it looks like he is on to something big, and even revolutionary. Evolution may indeed be governed much more by Law than by Chance, perhaps in ways Rhawn Joseph has suggested, perhaps in ways as yet undiscovered.

By the way, if as Behe implies, there is a designer or God, he is a very subtle God indeed, and a master scientist to boot. Because in the end, there is a scientific answer to everything.

Peter Breggin: If Peter Breggin is right, that leaves us with a dilemma: Is psychiatry an evil profession and if so, *why*? My sense is that this issue has to be put into some kind of perspective. No one can argue that there are not those who can not function in society, and for one reason or another, must be restrained. Also there is the question of people's right to use legal drugs if they so desire.

But surely the widespread use of psychiatric drugs and other harsh treatments does not reflect the relatively small number of individuals who really need them. Surely these drugs are overused and over prescribed. No doubt also psychiatry as a profession has too much legal authority, and no doubt the knowledge and expertise of psychiatrists is overrated both by the legal authorities and by the general public. Also, knowledge of the harmful and sometimes devastating effects of these drugs is downplayed in the professional literature or outright denied, and many patients are often kept in the dark until it is too late. Surely there is a danger here. This is a serious problem which must be rectified.

Patrick Michaels: No doubt Michaels' science would be thoroughly mainstream were it not for the fact that environmentalism is a key political issue for the Left. So widespread is their "scientific agenda" that the misconceptions surrounding this subject are taught throughout the land in our public schools and universities. This has been so for many years. In this sense and in this sense only, Michaels is fighting an uphill battle. In my opinion he and his co-workers have won the scientific fight hands down, but fighting an ideological battle is a different matter, and is much more difficult. One's opponents can be ruthless. It is a known historical fact that people will lie, cheat, steal, and even kill, given strong enough political motivations.

That is not to say, and we did not say, that there is no such thing as global warming or that there are no legitimate environmental issues, just that they are not as the mainstream model presents them.

Peter Duesberg: Dr. Duesberg has been all but defeated by the medical establishment. One can understand the motives of that establishment, but what of the rank and file medical doctors? One of the least emphasized factors—

and tragedies—concerning this issue, is the curious behavior of the medical professionals, especially the doctors. Surely a great number of them must have seen what we saw and could appreciate Duesberg's evidence. That there are thousands of good doctors with logical minds and good fundamentals goes without saying. Why then did they follow this obviously mistaken trend like lemmings walking off a cliff?

I have little doubt that Duesberg is right. Although I can not prove it, I would stake my own life on it, though not someone else's. That's how much I value the rational faculty we are all born with but which so many of us fail to use in certain cases. This apparently is one of them.

In any event, Duesberg has made his case. He got his message out, and no doubt he has saved lives. Only time will tell if he will be vindicated.

Petr Beckmann: Since Beckmann died some ten years ago, it is safe to say that he did not see Einstein's Theory of Relativity "placed on classical grounds," beside Galen and Aristotle and the rest. In other words he did not live to see that theory become antiquated and obsolete. He himself did not think that his "first attempt" would succeed, and admitted that he probably got some of the details wrong. But my sense is that his general thesis is correct, and that history will honor him as one of the pioneers who courageously challenged this bizarre theory.

Tom Van Flandern has also offered some insights, reported by Tom Bethel, into the incorrect predictions of Einstein's "Special Relativity" in this area.

Carver Mead: Dr. Mead, and again Beckmann gave us a good glimpse at what a deterministic, logically consistent, scientific theory of quantum electrodynamics might look like. It is a theory based on causality instead of chance, intended by Mead to replace the traditional statistical theories and weird implications of quantum mechanics. Although the statistical model has indeed "worked," which is one of the strongest arguments in its favor; this would not be the first time that an approximate theory was replaced by a more exact one, not to mention that the replacement theory provides more insight and is more "economical."

Van Flandern has again offered some other logical insights into some of the more absurd implications of QM in this area, and has taken the first tentative steps to combine his theory of pushing gravity with known electromagnetic phenomena.

Halton Arp: Bluntly, I think that the treatment which Arp has been given by mainstream astronomy has been nothing less than disgraceful. His observational evidence has been impeccable and must be correct. The theoretical framework supplied by others to explain his observations is quite plausible but needs further investigation. The only problem is that the mainstream is wedded to an untenable paradigm, namely the Big Bang theory, which states in effect that the universe was created from "nothing," and only very recently. Their motivations, in addition to devotion to the sainted Einstein, are the usual ones of preserving their petty prerogatives. The case of Arp and one or two others persuaded me of the necessity of writing this book.

Tom Van Flandern: Van Flandern is, in my opinion, one of the best minds of the group. Once again, many of his theories have been denied by the mainstream of astronomy, but are obviously correct. He has also tackled one of the most difficult and elusive of all issues in science, namely the *physical cause* of gravity. This has led him to "stick his neck out" in postulating a theory which is extremely difficult to prove. The future may very well honor him and a few others for one of the great discoveries in the history of science.

Van Flandern has been needlessly ridiculed for his advocacy of the artificiality of the Face on Mars. At this stage, that issue can not be settled. We must go there in person to see.

Zecharia Sitchin: Were it not for the Face, I would not have included Sitchin's theory in my survey. Sitchin is after all not a scientist. He is rather a scholar. But his Apocryphal theory nevertheless commands our attention in light of other facts we have discussed.

Robert Schoch: Like Michaels, Schoch's science should already be in the mainstream because of the clear-cut simplicity and recognized methods which led him to his conclusions. But then there is the political intrigue and unfortunate state of affairs in today's science, standing in the way again. Schoch's interpretation of the age of the Sphinx is obviously correct.

Rhawn Joseph: His theory of astrobiology and evolutionary metamorphosis is intriguing, but I don't have a good feeling for it yet on the question of its veracity. It's going to take further study. It's going to have to await the era of space travel to begin in earnest. It is also going to have to wait for the science of genetics to develop further. Then we will see.

THE FAILURE OF PHILOSOPHY

Science can not defend itself. Scientists are too busy dealing with facts. The integrated, common sense view of the world which allows science to exist in the first place has to come from somewhere. It has to come from philosophy. During the Enlightenment era, and building on Aristotle, it did come from philosophy. But modern philosophy, especially in the form of Kantian subjectivism, has according to Ayn Rand, abdicated its authority and its responsibilities, and as we have seen, it has helped produce many missteps in science. Today mainstream philosophy is no longer the seminal intellectual force it once was, and can no longer instruct the hard sciences. With respect to the hard sciences, this philosophy now supports the mainstream in most of the details and accepts its paradigms. It relegates to itself things like discussions of archaic concepts such as those of ancient Greece or of the Scholastics, or the dissection of arcane modern versions of logic and word parsing. It is fair to say, that modern philosophy in its present state, cannot solve this problem. Indeed, it may not even recognize that there is a problem, even though it was probably the cause of it a hundred years ago and still today.

THE "SHRINKING DOWN" OF AYN RAND'S PHILOSOPHY

Every generation has a kind of popular philosopher, or "man of the people" who common folks look to for answers to make sense of the world. It my generation, that "man" was a woman. Ayn Rand attempted to save philosophy from itself and in the process, to rescue science from its present crisis. But she barely began that project.

The followers of Rand's philosophy of Objectivism might have done better. Picking up where Rand left off, they might have done much to analyze and explain the current malaise in science. Instead, in an effort to be "accepted" by the academic world, they have chosen to embrace the mainstream in its investigations, with the exception of issues related to economics and a few other narrow philosophical fields where they have continued in the spirit of Randian polemics. Some Objectivist spokesmen have even denied the reality of Kuhn's concept of paradigms, discussed in briefly in these pages. So it has been left to scientists like Arp, Van Flandern and a few others, and science writers such as Tom Bethel, to attempt the task of dealing with the problems with science.

SOME CONSPIRACIES ARE REAL

Neither am I competent to attempt such a task. It is a job for some future scientist or philosopher who understands the problems involved and is aware of the implications of my rough sketch and broad outlines, but in the detail they properly deserve. I hope such a thinker exists or will soon appear on the scene. In the meantime it would be easy to do nothing but the bare minimum in life; to get along, to have some fun, and let events take their present course. That's what most people do, and there is nothing wrong with that. But history is determined by those who do more.

In these times as well as in the past, the truth is often defined as that which maintains the status quo and preserves the positions of those who control it. But that "truth" can be, and often is, challenged by men and women of vision and ability. Some conspiracies are indeed real, but you are free to do something to remedy the situation. Francis Bacon's famous dictum; "knowledge is power," is true.

Appendix A

New Science, Tesla, and Flimflam

PUTTING THINGS IN PERSPECTIVE

There is an underground movement afoot today called "New Science," which has barely been touched upon in this book. This may seem surprising since the topic of my book is precisely concerned with the closely related subject of unknown and / or suppressed scientific theories. Yet there are differences. I am aware of this new science and have had some exposure to it. One of the "fathers" of this movement is Nicola Tesla, who, despite his faults, has long been a hero of mine. But the essence and leitmotif of this new science is nonetheless at odds with my approach in certain significant ways. This appendix will deal with that issue.

Do I believe that UFOs exist, that anti-gravity technology is possible, that there are forces in nature that are as yet unrecognized by mainstream science? The answer is yes and no and maybe; it all depends. Do I think there are conspiracies to silence and suppress such facts? Same answer.

The point is—and this has been one of my main points throughout this book—that there are scientific ways to find out such things. There is good evidence to show that new, original, and most likely valid theories and facts exist, though they may be unknown to most people.

Moreover, these theories, and their advocates, are often fought "tooth and nail" by the mainstream of science and other mainstream institutions with a stake in the outcome, because if the new theories were to win out, some of them would have the ability to replace the existing paradigms. This has been shown in some detail in the cases of Halton Arp and Peter Duesberg, for example. It's true that this represents a kind of conspiracy, but probably not an organized one. But we've been through all that.

Another point of this book has been to show that in addition to ulterior political and financial motives, mainstream science is also strongly influenced by the dominant philosophical trends and ideas of our age. Indeed, one of the major reasons for the quick acceptance of certain new bizarre theories, outlandish though they might have been, is that they were in philosophical harmony with the prevailing wisdom of the day. I have tried to show, for example, that Einstein's Theory of Relativity, and the Copenhagen Interpretation of quantum mechanics would have been inconceivable without the relativistic and subjectivist philosophies in the air at the time, and still today for that matter. Today, new outlandish theories are being accepted by the mainstream for the same kinds of reasons.

To lesser degrees, sloppy reasoning and poor philosophical formulations, including faulty methodology, have made possible the success of other paradigms of science which the Apocryphal theories discussed in this book might credibly challenge. Examples here are Peter Breggin's challenge to modern bio-psychiatry, Peter Duesberg's challenge to "command science" in medicine, Michael Behe's and Rhawn Joseph's challenges to certain aspects of Darwinism, Tom Van Flandern's and Halton Arp's challenges to mainstream astronomy, Patrick Michaels' challenge to mainstream climatology, and Carver Mead's challenge to the "bewilderness" of Bohr's "principles."

But the main thrust of this book has not been philosophical or even conspiratorial; it has been practical; it has been rather to show that the challenging new theories are good scientific theories in their own right, and based on accepted scientific methods; and therefore they ought to be given a fair hearing. In the final analysis, the only argument that can win the day is the factual and scientific one.

THE NEW SCIENCE CONSPIRACY THEORY GENRE

So that is what I have tried to do. Can as much be said for the "New Science" enthusiasts of the "Conspiracy Theory Genre" available at your local bookstore. Tentatively, I say no. For example, a certain "inventor" may demonstrate a machine which supposedly draws current from the surrounding "space energy," and has an output which far exceeds the power consumption used to excite its circuits. But as we know from conventional physics, the axiomatic theory in such matters is the Law of the Conservation of Energy, expressed simply as; *Matter / Energy can be neither created nor destroyed*. In practical mechanical and electrical theory it is expressed this way; *Power IN equals Power OUT*, less Efficiency from heat loss, friction, and other mechanical or electrical losses.

In the New Science theory, the "input" comes from unseen and probably unknown "space energy" sources, and one can certainly accept this as a hypothesis. But that said, it then becomes a matter of proving it. At the risk of being accused of oversimplification, I have to say that any electrician could slap a wattmeter or ammeter on the input and output wires to tell you if this is indeed occurring.

Now it is certainly possible that the Earth's gravitational and/or magnetic fields are sources of this hypothetical energy. It is also possible that something even more elusive, such as Van Flandern's or LeSage's "gravitons" or Arp's "energy substrata field" (from which new galaxies seem to form), is the source. Perhaps these hypothetical particles, (and their associated forces), perhaps many orders of magnitude smaller than the smallest known subatomic particles, can someday be made to do useful work for mankind. This work may even include building perpetual motion machines and anti-gravity devices to power spaceships. I have high hopes that these things will someday be done. It is also possible that conventional electromagnetic forces, well known since the discoveries of Faraday, Maxwell, and others in the 19th century, will be shown to have new unknown properties which will allow this new science to work. It's all possible.

But the method of this book has been to take plausible if unconventional theories, and then to show good evidence, and then to critically evaluate the whole in a logical way, and let the reader decide. Not so for the typical book of the Tesla / New Science division of the Conspiracy Theory Genre. Old Tesla may be getting a bad rap here since he was a great theoretical electrician and pioneer, who gave us three-phase alternating current, modern power generation and transmission, was probably the first inventor of the radio, and much more. The fact that his memory is being ill-served by certain flim-flammers is unfortunate but probably true.

I'll give a brief example of what I mean by citing a recent book form this genre. The book is, *The Coming Energy Revolution*, by Jeane Manning (1996). Manning is a science enthusiast who is not herself a professional scientist. I don't think that fact alone disqualifies her. Still, her approach raises questions.

Manning surveys several leaders and "innovators" in the New Energy science world, and gives her take on the state of the art. Although certain oversimplifications can be explained by the fact that her book, like this one, is for the general reader and not for science specialists, I nevertheless question her for giving very few clear explanations on how the various inventions work. Add to this that almost all of her subjects were supposedly threatened and harassed by mysterious "conspirators" from government and / or big business. But unlike in my chapter on Duesberg, for example, the conspirators are never named.

Moreover, few of her subjects had inventions or discoveries which were "falsifiable" or testable. That is, they were not repeatable or verifiable by other scientists. Whereas, by way of making a contrast again, Duesberg has provided mountains of evidence and has many supporters and allies who are reputable and even famous scientists; Beckmann lays his formulas out for everyone to see and even publishes the criticisms of peers, and many of Van Flandern's theories are quite testable.

The same can not be said for many of the theoretical explanations behind the various inventions discussed by Manning; they are often vague and mysterious. Even certain known theoretical underpinnings which were once in vogue, such as the "aether" or "ether," of 19th century physics, (see Chapter 6), are poorly explained and understood as they relate to the new science. The reader can not begin to understand from reading Manning, how such an "ether" can have the properties the New Scientists claim for it. A reasonable objection might be that the reader must go to the principal sources to get detailed explanations—and I accept that. Still I am dubious because when I have previously read technical articles written by scientists in this genre, I was not convinced, and for the same reasons.

In the 19th century, when Koch and Pasteur were making their great discoveries in medicine, hucksters were selling "patent medicines," laced with opium and cocaine, from the backs of wagons. When Maxwell and Faraday were inventing electromagnetics, others were doing "magic" tricks or dreaming of "creating life" with the new science of electricity. So too now there is real science and there is flimflam. I'm not ready to say that all or even most of the inventors and theoreticians discussed by Manning are frauds and flimflam men, but their methods, and the tenor of her book leave me doubtful.

That said, there were scientists and discoveries covered in her book which I am confident are representative of good science. Moreover, I am sympathetic to her argument that there are economic and political forces which tend to stymie these valid new discoveries and inventions. I have documented as much in my book. Manning's best example of valid and potentially successful new science is "fuel-cell technology" for utilizing the element hydrogen as a safe and efficient source of energy. Another good example she gives is new environmentally friendly methodology for generating hydroelectric power. Still I am skeptical of many of her subjects' theories, such as those that claim anti-gravity properties, and those which claim to draw energy from the ether. Until I see reputable verification of these claims, I will withhold my approval. A good forum exists for reviewing such claims from a strictly scientific point of view but without the constraints of mainstream preconceptions. This is Tom Van Flandern's Meta Research web site and message board on the Internet, previously mentioned in Chapter 10.

Appendix B

Good Nutrition and Health Hoaxes

WEIGHTY MATTERS

In the Introduction to this book I wrote that it was not my intention to deal with diet fads or junk science, but with "weightier matters," and I have tried to do so. But since that time, I have come to realize that the matter of personal health is indeed a life or death issue for many people—including me. Diet and nutrition come under this general heading, as do medicine and health care. This appendix is on a topic which is one small aspect of that larger subject. It will be my final installment to this survey of Apocryphal science. The topic is fat and cholesterol, related nutritional factors, and associated health risks, including coronary heart disease, arteriosclerosis, and heart attacks.

I mentioned also in the Introduction that *The Omega Diet,* by Artemis P. Simopoulos and Jo Robinson (1999), was the best diet book. But it is much more than that. It is a reasoned, scientific study of the effects of nutrition on health and disease prevention. It utilizes the latest research and politely dissents from the mainstream paradigm in many respects. An important aspect of Simopoulos's book is that it challenges "the cholesterol myth." But it does so only incidentally, as a side issue to the author's main area of interest.

The Omega Diet centers on the principle author, Simopoulos's scientific opinion that it is not high fat intake *per se* that is the significant causal factor in coronary heart disease and other illnesses, but *the kinds of fats consumed.* Citing the traditional diet of the Greek island of Crete and several recent scientific studies, she offers convincing evidence that a diet high in "Omega 3" fatty acids, (found in tuna, salmon, spinach, purslane, walnuts, and canola oil; among other things), is essential to good health. A landmark study in this regard was, "The Lyon Heart Study," conducted by researchers S. Renaud and M. de Lorgereril (Simopoulos and Robinson 1999, 7).

The consumption of Omega 3 fatty acids must be balanced with other kinds of essential fats consumed, especially the Omega 6 fatty acids, which is the type most consumed in the typical American diet. Omega 6 fatty acids are found in most vegetable oils, such as corn oil and soybean oil, and in margarine, They are also found in most commercially packaged foods, especially, but not limited to, baked goods and snack foods.

In a closely related vein, she cites the dangers of the once lauded and promoted as heart healthy, "polyunsaturated" fats and oils. She contrasts these to the health benefits of "monounsaturated" oils such as are found in olive oil, canola oil, and to a lesser degree, soybean oil. When polyunsaturated fats and oils contain high proportions of Omega 6 oils, they are more harmful, but combined with Omega 3 oils they are more beneficial. Both types of these natural fatty acids are essential nutrients in that the human body can not manufacture them. *It is the proper ratio of Omega 6 to Omega 3 oils that is critical.* (The ratio should be at least 4 : 1. The details of her theory are beyond the scope of this brief synopsis. I refer the reader to her excellent book for a more in-depth explanation. It also has many delicious recipes).

Most dangerous of all in Dr. Simopoulos's pantheon of fats and oils is *not* the "usual suspect," which is saturated fat—found in meat, milk, and eggs, although she does advise moderation in the consumption of saturated fats. It is instead the "trans-fatty acids," which are most harmful. These are fatty acids produced by heating and reheating polyunsaturated oils such as in deep frying, or by "hydrogenation." Hydrogenation is a commercial process whereby oils are solidified to make vegetable shortening or margarine. Hydrogenated oils are also a ubiquitous ingredient in many packaged food products, such as breakfast cereals, breads, crackers, cookies, donuts, and much more. Trans-fatty acids are linked to cancers, heart disease, and other illnesses.

All of this underscores the nutritional deficiencies of the modern American diet. It is the reason she recommends a diet favored by Mediterranean peoples, rich in dark green vegetables and including lots of fresh fish, nuts, fruits, and a balance of fats and oils as mentioned above. She also recommends seven servings of fruits and vegetables daily, and she favors olive oil and canola oil over other vegetable oils as the table oils in your home.

The Omega Diet and its author could have easily been included in my survey of Apocryphal science in a more detailed analysis. Here is a scientific theory that is critical of the current paradigm on fat, cholesterol, and nutrition in general. She criticizes for example, the so-called "Prudent Diet" of the American Heart Association which promotes a low-fat diet, but is slow to recognize recent research proving that it is *the kind of fat* that is critical. The Prudent Diet also continues to promote unhealthy polyunsaturated, and hydrogenated fats such as margarine, and artificial "diet foods" high in refined sugars and carbohydrates, and various chemicals of questionable benefit, such as artificial sweeteners.

In addition to the above, Simopoulos makes several other extraordinary claims; all backed by scientific studies, about the health benefits of a diet high in Omega 3 fats. Among these are reduced risks for several types of cancer, adult onslaught or "type II" diabetes, clinical depression, and more. She does not recommend her diet as a substitute for medical treatment, but as a preventative measure and nutrition-based lifestyle. This is a book well worth reading and discussing with your physician.

I regret not including Simopoulos as one of my Principals. My main reason for not doing so was that I didn't think diet and nutrition were serious enough scientific subjects. In light of recent reading however, and recent losses of family members to heart disease, I have reconsidered, and decided that this is an important issue and should be discussed.

HOAX 101

In addition to *The Omega Diet*, another book I read recently has prompted me to take one more plunge into what I consider to be yet another "good conspiracy theory." This book, *The Cholesterol Hoax / 101 + Lies*, by Sheldon Zerden (1997), says that the 50 year old paradigm which claims that high cholesterol is the major causative factor in heart disease—is wrong.

Any new Apocryphal theory, when first encountered, may easily strike one as "crackpot," and engender skepticism. This is especially true when the new theory opposes a widespread and commonly held view that affects almost everyone's lives. More than that, it may have a direct bearing on whether one lives or dies.

Everyone wants to trust his or her doctor. When your doctor tells you that your cholesterol is high, and that there is good medication available to lower it, you follow your doctor's orders. It's hard to stand on one's own independent judgment.

This is all true—to a point. But doctors are people too. They in turn take orders from those higher up in their fields. They don't want to be sued for malpractice. They don't want to lose their license to practice medicine. If the medical establishment, including the American Medical Association and the American Heart Association, says that even moderately high cholesterol is dangerous, that cholesterol-lowering drugs are safe, and that an extremely low-fat and high-carbohydrate diet is nutritionally proper, who is your doctor to argue? Who are you to argue? But that doesn't make you any less of a "failed trial" if they are wrong.

You, the reader, can think. You can decide for yourself. Here is more information to help you to do so. Again, as before, I don't say that you should go against your doctor's advice. Only that, if interested, you should study the subject further, and then decide.

CHOLESTEROL

In his book, Sheldon Zerden, who is also a writer of books on the stock market, makes a fairly reasonable and convincing case, but not thoroughly convincing one, to prove his paradigm-busting theory. Following is a sampling taken from his list of "cholesterol lies." In the interest of brevity, the "lies" will be converted to statements which Zerden would consider factual and confirmed by scientific studies, many of which are cited by that author.

1. Dietary cholesterol does not affect blood (serum) cholesterol (Zerden 1997, 1).
2. According to researchers, the intake of saturated fat and overall cholesterol has no effect on serum cholesterol (4).
3. From 1900 to 1965, heart attacks of Americans increased dramatically, but cholesterol levels remained constant (6).
4. Polyunsaturated fats were found to cause cancer in laboratory animals (9).
5. Cholesterol-lowering drugs do not increase life expectancy, tests show, because subjects had increased deaths from "other causes," notably, cancer (10, 28, 82, and 109).
6. 33 studies in 30 years have failed to prove that high saturated fat and high cholesterol cause heart disease. The "diet-heart hypothesis" is not a sound one (11).
7. Several examples of ethnic populations or indigenous peoples who consume a high-fat diet but have a low incidence of heart disease have been shown; e.g., rural Rumanians (1), the Maasai of Kenya (11), the French (16).
8. Several examples of peoples (nations or ethnic groups), who switched from saturated fat diets to polyunsaturated fats revealed an increase in the incidence of heart disease; e.g., the bedouin of Israel (13), Norway after WWII (14), and the USA during the first half of the twentieth century (22, 68).
9. Trans-fatty acids found in margarine, and high-carbohydrate diets, both recommended by the Prudent Diet, raise LDL (bad cholesterol), and lower HDL, (good cholesterol), (13, 38).

10. The death rate from heart disease of vegetarians is comparable to that of non-vegetarians. But vegetarians have a higher death rate from all causes (16).
11. The Prudent Diet, which recommends a low-fat diet, including low-fat or skim milk, few or no eggs, vegetable oils, peanut butter, margarine, and potato chips; subjects children to "nutritional dwarfing" (17).
12. Egg yokes, shunned by the Prudent Diet, are good for you. They contain many essential nutrients and taste good too (19).
13. The Framington study, (a landmark study disproving the orthodox "cholesterol myth"), shows no increased death rates with increased serum cholesterol (20).
14. Animal fat consumption *decreased* during the heart disease epidemic of 1920—1989 (22).
15. Most people with coronary heart disease (CHD) have *low or moderate* serum cholesterol levels (24).
16. No relationship was found between animal fat consumed and serum cholesterol (25).
17. Red meat has less cholesterol than fish and the same as chicken—two foods favored by the Prudent Diet (26).
18. Studies citing Greenland Eskimos that claim that their diet which is high in fatty fish is heart-healthy, are faulty. For one thing, few Eskimos live beyond fifty years, the age at which most heart disease sets in. For another, fish oil supplements were not shown to lower CHD rates (28).
19. Cholesterol is an essential substance in the body. If you don't consume enough, and eat a high-carbohydrate diet instead, your body makes its own cholesterol, and it makes the bad kind, LDL (36, 38).

MULTIPLE FACTORS

Although Zerden goes to considerable lengths to disprove the "cholesterol myth," he doesn't do much to clearly state what kind of a diet is good for you, as does Simopoulos. Neither does he always make his case in an unambiguous way. For example, he admits implicitly, that one kind of cholesterol (LDL) is harmful (Zerden 1997, 13, 120). So in a way, some of his arguments amount to "beating a dead horse," because mainstream nutritional science has moved beyond the outmoded claim that "all cholesterol is bad." On the other hand, much of Zerden's case remains compelling and reasonable—especially the parts linking polyunsaturates and trans-fatty acids to disease—as long as the reader takes it with a healthy dose of skepticism. Again, the interested reader should study the subject further, and consult a physician who is well-versed in nutritional science. Nevertheless, Zerden, and other contributing authors of his book, make some salient points.

Zerden seems to be saying that saturated fats are okay, at least in amounts consumed up till the beginning of the twentieth century. But a scientific case can be made for noting that the fat contents of red meats such as beef, has increased due to modern agricultural methods (Simopoulos and Robinson 1999). It is also obvious that people are more sedentary than they were in the past, and more obese. These are obviously contributing factors. Therefore it would not be reasonable to imply that we can consume unlimited amounts of saturated fats and cholesterol with impunity. To be fair, Zerden does not say this, but the implication is there.

One contributing author, W. M. Levin, opines that it is not saturated fat in itself which is dangerous, but in effect it is "tainted meat" that is the culprit. This opinion stems from the fact that "chemicals" are used in the agricultural process at all stages; from fertilizers to insecticides to the herbicides used in the farming of feed corn which is the major staple of beef cattle. The process continues with the injection of hormones to produce fatter cows, (Zerden 1997, 124–26). Add to this the fact that most domesticated cattle no longer "free range," and consume a less nutritious diet (Simopoulos and Robinson 1999). The result is beef which is less nutritious than beef may have been in the past—and perhaps more harmful. This kind of evidence is not specific nor is it definitive, but it should encourage a prudent consumer to limit his or her intake of beef and similar products.

Simopoulos makes the point that evolutionarily speaking, man ate very little grain foods before the advent of the agricultural revolution 10,000 years ago. Grains are grasses; in nature the seeds are too small to provide humans with much food. But once man learned to farm grain and make bread, he altered his genetically natural diet (Simopoulos and Robinson 1999).

I know of no evidence that whole grains are harmful if consumed moderately, since they are known to be nutritious foods. But I have little doubt that, as Zerden points out, (1997, 111), *refined* sugars, and flour can't be good, for a number of reasons. That they contribute to the myriad of diseases which appeared in the twentieth century seems fairly clear.

Is alcohol consumption good or bad for you? Many of us are familiar with the anomaly of the French who, while eating copious amounts of cheese, and drinking wine with every meal, have less heart disease than Americans. My brother Richard points out the following: "There've been many anecdotal tales of Italians in northern Italy living to 100 while drinking two glasses of wine a day." Richard also points out other evidence which Zerden provides to support this position: "No one advocates alcoholism," Zerden writes, "but with the billions spent on research, one physician aptly put it, 'Americans should be told that not drinking alcohol is a "cardiac risk factor." And, "The harmful effects of alcohol occur only at high doses" (Zerden 1997, 33). Elsewhere he states: "Two between-nation studies showed that the more alcohol consumed by a nation, the lower the coronary heart disease rates, less atherosclerosis and larger diameter arteries were found in the drinkers" (27).

On the other hand, one should take such information with a grain of salt. One reason is that no one knows exactly what constitutes "alcoholism," and at what level of consumption its degenerative effects begin. Simopoulos cites a study in this connection indicating that a high percentage of heavy drinkers suffer from depression, and worse, that alcohol is one of the few known substances that leaches DHA (a type of Omega 3 fatty acid which is an essential part of brain tissue) from the brain (Simopoulos and Robinson 1999, 93). All of these matters are worth considering for anyone who wants to eat well and remain healthy.

References

Arp, Halton. 1987. *Quasars, Redshifts and Controversies*. Interstellar Media, Berkeley, Cal.

———. 1996. *Seeing Red: Redshifts, Cosmology and Academic Science*. Apeiron, Montreal.

Beckmann, Petr. 1987. *Einstein Plus Two*. The Golem Press, Boulder, Colorado.

Bethel, Tom. 1993. "Doubting Dada Physics," *The American Spectator*, (August). R. E. Tyrrell, ed. Arlington, Virginia.

———. 1993. Correspondence, "Dada Physics," (October).

———. 1999. "Rethinking Relativity," (April).

———. 1999. Correspondence, "Rethinking Relativity," (June).

Behe, Michael J. 1996. *Darwin's Black Box*. The Free Press, New York.

Breggin, Peter R. 1991. *Toxic Psychiatry*. St Martin's Press, New York.

———. 1995. *Talking Back to Prozac*. St Martins.

———. 2000. *Your Drug May Be Your Problem: How and Why to Stop Taking Psychiatric Medications*. Perseus, Cambridge, Mass.

———. 2001. *The Anti-Depressant Fact Book*. Perseus.

———. 2001. *Talking Back to Ritalin*. Perseus.

———. 2003/2004. "Suicidality, violence and mania caused by selective serotonin reuptake inhibitors (SSRIs): A review and analysis," *International Journal of Risk and Safety in Medicine* 16, 31-49. IOS Press. http://www.breggin.com/

Copleston, Frederick. 1993. *A History of Philosophy*, Vol. I: *Greece and Rome*. Image Books, New York.

———. 1994. Vol. VI: *Modern Philosophy: From the French Enlightenment to Kant*.

Courtois, Stephane, et. al. 1999. *The Black Book of Communism: Crimes, Terror, Repression*. Harvard University Press.

Crewdson, John. 2002. *Science Fictions / A Scientific Mystery, A Massive Cover-up, and the Dark Legacy of Robert Gallo*. Little Brown & Co., Boston.

Dawkins, Richard. 1996. *The Blind Watchmaker*. W. W. Norton, New York.

Duesberg, Peter H. 1996. *Inventing the AIDS Virus*. Regnery, Washington, D. C.

——. 2000. "The African Aids Epidemic: New and Contagious—or—Old Under a New Name?" http://www.duesberg.com/subject/africa2.html

Edwards, Matthew R., ed. 2002. *Pushing Gravity*. Apeiron, Montreal.

Einstein, Albert. 1936. "Physics and Reality," "What is the Theory of Relativity?" *Ideas and Opinions*. Bonanza Books, New York

Fortey, Richard. 1997. *Life / A Natural History*. Vintage Books, New York.

Gamow, George. 1961. *The Great Physicists from Galileo to Einstein*. Dover, New York.

Gribbin, John. 1984. *In Search of Schrödinger's Cat*. Bantam Books, New York.

——. 1995. *Schrödinger's Kittens and the Search for Reality*. Little Brown, Boston.

Hall, A. Rupert. 1981. *From Galileo to Newton*. Dover, New York.

Heyek, Friedrich A. 1954. *Capitalism and the Historians*. University of Chicago Press, Chicago.

Hellander, Martha. 2000. *Newsletter of the Child & Adolescent Bipolar Foundation*, (November, 21). http://www.eletra.com/cabf/.

Hoagland, Richard. C. 1996. *The Monuments of Mars*. North Atlantic Books, Berkeley, Cal.

Horgan, John. 1996. *The End of Science*. Addison-Wesley, New York.

Horowitz, David. 1998. *The Politics of Bad Faith*. The Free Press, New York.

Jaakkola, Toivo. 2002. "Action-at-a-Distance and Local Action in Gravitation." *Pushing Gravity*, Mathew Edwards, ed., Apeiron, Montreal.

Johnson, Paul. 1994. *Modern Times / The World from the Twenties to the Eighties*, Harper and Row, New York.

Joseph, Rhawn. 2001. *Astrobiology, The Origin of Life and the Death of Darwinism*. University Press California, San Jose, Cal.

Kant, Immanuel. 1950. *Prolegomena to Any Future Metaphysics*, The Liberal Arts Press, New York.

Kauffman, Stuart. 1995. *At Home in the Universe / The Search for the Laws of Self-Organization and Complexity*. Oxford University Press, New York.

Kuhn, Thomas. S. 1970. *The Structure of Scientific Revolutions*. University of Chicago Press, Chicago.

Leakey, Richard. 1992. *Origins Reconsidered / In Search of What Makes Us Human*. Anchor Books, New York.

Manning, Jeane. 1996. *The Coming Energy Revolution*. Avery Publishing, Garden City Park, New York.

Martins, Roberto de Andrade. 2002. "Majorana's Experiments on Gravitational Absorption." *Pushing Gravity*, Mathew Edwards, ed., Apeiron, Montreal.

Mead, Carver. 2000. *Collective Electrodynamics, Quantum Foundations of Electromagnetism*. The MIT Press, Cambridge, Mass.

Meta Research website. http://metaresearch.org/.

Michaels, Patrick. J. 1992. *Sound and Fury*. CATO Institute, Washington, D. C.

Michaels, Patrick, and Robert Balling, Jr. 2000. *The Satanic Gases*. CATO Institute.

Morris, Desmond. 1999. *The Naked Ape*. Delta Books, New York.

Rae, Alastair. 2000. *Quantum Physics: Illusion or Reality?* Cambridge University Press, Cambridge, UK.

Rand, Ayn. 1952. *The Fountainhead*, Signet, New York.

———. 1967. *Capitalism the Unknown Ideal*. Signet, New York.

———. 1992. "Galt's Speech," *Atlas Shrugged*, Dutton, New York

Robison, Keith. 1996. "Darwin's Black Box, Irreducible Complexity or Irreproducible Irreducibility?" The Talk.Origins Archive, http://www.talkorigins.org/.

Root-Bernstein, Robert. 1993. *Rethinking AIDS*. The Free Press, New York.

Rustow, Alexander. 1980. *Freedom and Domination: A Historical Critique of Civilization*. Princeton Univ. Press, Princeton, New Jersey.

Sagan, Carl. 1980. *Cosmos*. Random House, New York.

Shenton, Joan. 1998. *Positively False/ Exposing the Myths around HIV and AIDS*. Tauris Pub., London.

Simopoulos, Artemis P., and Jo Robinson. 1999. *The Omega Diet*. Harper Perennial, New York.

Szasz, Thomas. S. 1974. *The Myth of Mental Illness*. Harper and Row, New York.

Sitchin, Zecharia. 1978. *The 12th Planet*. Avon Books, New York.

Shamos, Morris. H. 1987. *Great Experiments in Physics*. Dover, New York.

Schoch, Robert M. 1999. *Voices of the Rocks*. Harmony Books, New York.

———. 2003. *Voyages of the Pyramid Builders*. Tarcher / Putnam, New York.

Van Flandern, Tom. 1993. Dark Matter Missing Planets & New Comets, North Atlantic Books, Berkeley, Cal.

———. 2001. "Preliminary Analysis of 2001 April 8 Cydonia Face Image." *Meta Research Bulletin*, (June). Chevy Chase, Maryland.

———. 2002. "Gravity." *Pushing Gravity,* Mathew Edwards, ed., Apeiron, Montreal.

———. 2003. "21st Century Gravity." *Meta Research Bulletin*, (June).

———. 2003. "The Structure of Matter in the Meta Model." *MRB*, (December).

Von Daniken, Erich. 1970. *Chariots of the Gods?* Bantam Books, New York.

Von Mises, Ludwig. 1963. *Human Action / a Treatise on Economics*, Yale University Press.

Windelband, Wilhelm. 1950. *A History of Philosophy*, Vol. II, Harper and Row, New York.

Zerden, Sheldon. 1997. *The Cholesterol Hoax / 101 + Lies*, Bridger House Publishers, Carson City, Nevada.

Index